T0350079

Education and Training for the Oil and Gas Industry

Localising Oil and Gas Operations

Education and Training for the Oil and Gas Industry

Localising Oil and Gas Operations

Volume 4

Jim Playfoot

Simon Augustus

Phil Andrews

ELSEVIER

AMSTERDAM • BOSTON • HEIDELBERG • LONDON
NEW YORK • OXFORD • PARIS • SAN DIEGO
SAN FRANCISCO • SINGAPORE • SYDNEY • TOKYO

Elsevier
Radarweg 29, PO Box 211, 1000 AE Amsterdam, Netherlands
The Boulevard, Langford Lane, Kidlington, Oxford OX5 1GB, United Kingdom
50 Hampshire Street, 5th Floor, Cambridge, MA 02139, United States

Notices
Knowledge and best practice in this field are constantly changing. As new research and experience
broaden our understanding, changes in research methods, professional practices, or medical
treatment may become necessary.

Practitioners and researchers must always rely on their own experience and knowledge in
evaluating and using any information, methods, compounds, or experiments described herein.
In using such information or methods they should be mindful of their own safety and the safety
of others, including parties for whom they have a professional responsibility.

To the fullest extent of the law, neither the Publisher nor the authors, contributors, or editors,
assume any liability for any injury and/or damage to persons or property as a matter of products
liability, negligence or otherwise, or from any use or operation of any methods, products,
instructions, or ideas contained in the material herein.

Library of Congress Cataloging-in-Publication Data
A catalog record for this book is available from the Library of Congress

British Library Cataloguing-in-Publication Data
A catalogue record for this book is available from the British Library

ISBN: 978-0-12-800980-2

For information on all Elsevier publications
visit our website at https://www.elsevier.com/

  Working together
to grow libraries in
developing countries

www.elsevier.com • www.bookaid.org

Publisher: Candice Janco
Acquisition Editor: Amy Shapiro
Editorial Project Manager: Tasha Frank
Production Project Manager: Paul Prasad Chandramohan
Designer: Mark Rogers

Typeset by TNQ Books and Journals

Contents

About the Authors

JIM PLAYFOOT

Jim is a consultant, researcher and writer working in the field of education and skills development. He is Founder and Managing Director of London-based education consultancy White Loop.

Jim's work over the last 10 years has focussed on understanding the dynamic between education and employment, exploring the challenges of how we prepare young people for the 21st century and developing new thinking around how education can have real impact on wellbeing and quality of life. Jim's work is built around a deep understanding of how people learn allied to an ability to engage, analyse and interpret evidence and opinion and produce outputs that are compelling and accessible. He has worked with partners and collaborators in more than 20 countries around the world and is increasingly focused on exploring the potential for education as a means of promoting development objectives in Africa.

In 2011, he was approached by Getenergy Events to become involved in developing their research and intelligence function. In 2014, Jim was part of the team that established Getenergy Intelligence and assumed the role of Managing Director, a position that has seen him lead on the authoring of all four of the Getenergy Guides volumes.

SIMON AUGUSTUS

Simon is a researcher, analyst and strategist for the energy sector as well as other key sectors of the economy. He has worked with numerous high-profile clients from across the globe, assisting them to strategically channel investment, influence strategic initiatives and undertake numerous key activities and business-building exercises in new and developing markets.

Simon has spent most of his career researching and analysing the oil and gas industry as well as other high-reliability sectors of the economy. Simon works with a range of international clients, including the governments of Turkey, Egypt and Kurdistan/Iraq. His work today is particularly concerned with exploring the dynamic between the oil and gas industry and the wider geopolitical context alongside a focus on economic development and sustainable growth. Simon's articles are regularly published and you can find them on energy, renewables and industrial gases online or in print.

PHIL ANDREWS

Phil Andrews is one-half of the partnership which founded Getenergy back in 2003. The company has been defined by its entrepreneurial spirit and a profound dedication to enhancing the relationship between industry and education ever since.

Phil is the CEO of Getenergy Field-Ready and is currently working to support colleges and technical universities around the globe in deploying the Field-Ready system to produce highly employable staff for the oil, gas and wider energy industry.

Phil has just been awarded a master's degree from the University of Warwick following the University's decision to grant him a prestigious scholarship on their Global Energy MBA programme.

Preface

Welcome to the fourth volume of the *Getenergy Guides* to *Education and Training for Oil and Gas*. When we started writing this series, we had a clear aim – to create a focal point for the discussion around how we best educate and train the oil and gas workforce globally. We set out to discover models of best practice, to research and analyse case studies from around the world, to speak with those who had most direct experience of facing down the challenges of skills and workforce development and to make a contribution to the improvement of competency, safety and productivity across our industry. For those of you who have read the preceding three volumes, we hope you will agree that we have achieved many, and perhaps even all, of these objectives.

In our first volume, we set out ideas that, then and now, shaped our approach to the world within which we work. Specifically, our stated aim was to illuminate the interactions between three seemingly very different and often complex subject areas: energy, education and economy. We put forward the view that developing a greater understanding of how these subjects interact can make the business of extracting hydrocarbon resources better for the citizens of oil- and gas-producing countries, better for the companies doing the extraction, better for the governments who licence the exploration blocks in the first place, and better for national companies and their supply chains. We also noted that countries that have oil and gas resources but lack the expertise or experience to develop and produce these resources must necessarily rely on foreign companies to undertake the complex and often challenging task of exploiting those resources. And that this inevitably hampers the long-term economic and societal benefits that a country should enjoy through the gift of abundant natural resources. In the subsequent case studies explored across volumes 1, 2 and 3, we highlighted a raft of compelling stories – from the life of Enrico Mattei and the rise of ENI in the 1950s to the attempts by Tullow Oil to build localised training capacity in Ghana – that we believed illuminated how oil companies, governments and educators could support the professional development of competent, safe and productive employees.

Our hope and ambition was that these books would be referenced in countries around the world where the business of oil and gas exploration and production generates an inevitable – and significant – demand for one particular resource: competent, committed people. The context we were dealing with when we set out was characterised by a post-Macondo focus on safety, the rising importance of highly technical forms of extraction and production, an ageing (and retiring) workforce – at least in some parts of the world – and the emergence of a group of new or evolving energy nations that typically lacked the skills, experience and expertise to fully exploit their own hydrocarbon resources.

Since publishing our first book in this series in 2013, the context is both the same and very different. The industry is now in a permanent state of concern and focus over safety with the cost of Macondo still being counted and a recognition that the environmental, human and economic impact of accidents and disasters of that kind is, today, simply unacceptable. We continue to face ever-greater technical challenges with an increasing focus on accessing harder-to-reach hydrocarbons through deepwater drilling, enhanced oil recovery and fracking (to name but three). There persists a challenge around what the Americans call the 'big crew change' (although it is only a challenge in some parts of the oil- and gas-producing world). Certainly, we have seen developed energy nations largely fail to find the right ways of capturing, leveraging or distilling the experience of seasoned staff to support the development of the next generation of oil and gas worker. And we are certainly seeing the geological sands shifting in terms of where future oil and gas production is moving. By some estimates, around 90% of all oil and gas globally will be produced in emerging or developing economies by 2030 with Africa and South America leading the way in terms of new discoveries and untapped reserves. All of these factors continue to exert a significant influence on how we think about education and training and how we approach the challenges of developing competent, committed workers for the 21st century oil and gas industry.

However, these factors have, since the middle of 2014, played second fiddle to the dominant narrative across the oil and gas sector, that of a falling (or, perhaps, 'fallen') oil price and a comparable fall in the price of gas. At the time of writing, we are 2 years into the current 'oil crisis'. Arguably, it is starting to look more like a price adjustment than anything one could confidently term a 'crisis'. A crisis is, by its nature, temporary. This, it seems, is not. Whilst there are some who believe that the heady days of $100-plus a barrel are somewhere just around the corner – let us call them 'optimists' – much analysis has focused on the mismatch between rising global supply and falling demand, the shift in the political narrative supported by the Paris Climate Change Summit towards clean, renewable sources of energy and the realities that these two factors point to: something between $30 and $50 a barrel is quite possibly where oil prices will now rest for the foreseeable future. The message is clear: lower oil and gas prices are here to stay, and we must adapt to this reality.

This has profound implications for anyone and everyone involved in any aspect of the oil and gas business (including those engaged in the provision of education and training). First, it means that the viability of oil and gas projects is under ever more scrutiny. Simply put, the level of investment in new projects has fallen off dramatically and those investments that are continuing to receive funding tend to be in areas and regions where oil or gas production costs are low enough to ensure that money can still be made. This means that the relative commercial attractiveness of an oil or gas opportunity will determine whether international operators show any interest in bidding for concessions. This may also be the defining factor in any decision that is taken by governments and national oil companies in relation to new exploration and production projects. And that attractiveness relates both to the potential reserves and to the cost of exploiting those reserves. Countries that manage to create an operating environment that helps international producers to keep costs down will inevitably see greater inward investments and more new projects getting a green light.

At a more systemic level, the 'new normal' oil price has, and will continue to, reduced the revenues that governments can rely on now and in the future. We have already seen many countries that are highly dependent on oil or gas revenues struggle economically (and also politically) within this new price context (Nigeria, Azerbaijan and Kazakhstan spring to mind, although there are many others). Within this economic climate, ideas of ownership and sovereignty bubble up to the surface, and the need to connect local communities and local citizens to the natural resources that surround them becomes even more pressing. For operating companies, the resource price has simply changed the equation in terms of revenues, profits and, as a consequence, business strategy and investments. Many exploration projects have now been cancelled or, at best, shelved and the focus for a majority of operators is now to get the most out of existing assets and only to invest in parts of the world where the cost of operation is lowest and the yield highest. Let us make no mistake – these companies are still making money. However, the days of plenty may be over and a new frugality is currently shaping the mood across the industry.

The oil price has also had a profound effect on the dynamics of education and training across the sector. For those who supply education and training services to the industry – the colleges, universities, private training providers, technology companies and so on – the pressures are clear. Oil, gas and service companies – both international and national – now want better results, in shorter timeframes, for less money. The need to educate and train has not disappeared but recruitment levels are down and many companies are squeezing training budgets as far as they can. That said, the drain of experienced talent from some parts of the industry continues, and the emergence of a new breed of energy-producing nations across the developing world is generating a demand for skilled, competent people at every level. Most significantly, the pressures on cost are starting to shed light on the way in which oil and gas companies typically resource their projects, particularly in less developed parts of the world. There is now clear recognition that the expatriate-driven model of staffing – whereby a highly experienced and mobile global workforce are deployed (at considerable expense) as and where they are needed – is now looking somewhat unfit for purpose. The drive towards a new model – one that reduces costs, increases local revenues and generates value to governments, communities and citizens – is gathering momentum. We call this model 'localisation'.

It is our belief that localisation – by which we mean the sustainable and profitable functioning of international oil and gas operators within a local environment – is now the only effective model there is for finding, extracting and processing hydrocarbons around the world. The current model is outdated, ethically questionable and economically unworkable. The economic and industrial colonisation pursued – implicitly or explicitly – by international oil companies was always wasteful and morally suspect. But now, under persistently challenging economic circumstances, the industry as a whole has been forced to reevaluate, recalculate and restructure. Through the first three volumes of the *Getenergy Guides* series, we told many stories – stories of partnership, collaboration, endeavour, ambition, achievement – but too often these efforts have been in isolation, beacons of hope in an industry characterised by waste and inefficiency. When we sat down to write our fourth and final volume in the series, we decided that we needed to address the industry's current wasteful practices and delineate a new, sustainable model of operation.

So here it is – a vision for how the upstream oil and gas industry now needs to operate and a set of ideas, concepts, mechanisms and methods for making it happen. This book not only reflects everything we have learnt through our thorough and extensive research over the course of writing the *Getenergy Guides* series but is also the culmination of every meeting, event, project and conversation we have had since we started *Getenergy* back in 2004. Put simply, this book sets out what we believe, why we believe it and what we think should be done about it. For our part, we are placing the concept of localisation at the heart of everything we do – from our global events and national summits to our own Field-Ready employability programmes to the research we undertake and the consultancy services we provide to our clients. We are working today with governments, ministries, oil companies, services companies, educators, trainers and others to create a clear vision of education and training that supports the development of local talents and skills in countries with oil and gas resources. We hope that you will join us in this most vital of endeavours.

Jim Playfoot, Phil Andrews and Simon Augustus

Acknowledgements

We would like to thank the team that has helped us produce this book:

- Getenergy's Cofounder Peter Mackenzie Smith who read and reviewed these pages.
- The Getenergy team in London: Helen Jones, Jack Pegram, Richard Harmon, Conchi Perez, Virginia Baker, James Fox, Dominique Broomes and Shalil Gunputh.

We would also like to extend profound thanks to all the contributors who have given up their time and shared their stories with us for this book and for the preceding three volumes.

About Getenergy Intelligence

Getenergy Intelligence was established in 2014 to support the aims of the wider Getenergy group of companies, namely to bring new intelligence to the development of skills and competence for the energy industry. This mission is underpinned by the belief that countries that have abundant natural resources—and the operating and service companies that explore for and produce those resources—have a specific set of responsibilities that include the following:

- To grow the energy sector, the energy supply chain and related industries in a way that brings opportunity to citizens and enables those citizens to play an active role in this economic and industrial growth;
- To achieve this through helping citizens to develop the knowledge, competencies and behaviours that are demanded by employers and that will be transferable to other sectors;
- To create this opportunity for citizens by building a comprehensive, sustainable and accessible education and training system in partnership with industry;
- To build this education system locally and involving local education and training providers in a partnership of equals; and
- To have a long-term vision around workforce and skills development that sees beyond the immediate skills needs of the energy sector and envisions a future economy that is diverse, dynamic and competitive.

Getenergy Intelligence is committed to working with governments, educators, employers and others to create debate, stimulate engagement and, through this process, capture the stories, ideas, opinions, data and case studies that emerge from these activities. The desire to capture, distil and share global good practice is made real through the publication of the Getenergy Guides series and reflects our mission to support and promote the development of effective education and training in hydrocarbon-producing countries.

Chapter 1

The Rationale for Localisation

Chapter Outline

1.1 THE CONTEXT FOR LOCALISATION: A CENTURY IN THE MAKING

The prevailing business model that underpins oil and gas operations around the world has remained largely unchanged since the first half of the 20th century. We have seen the rise of the international oil company (IOC) and the emergence of a group of organisations that, collectively, would come to dominate the global industrial landscape for generations. As the world's demand for oil (and, latterly, gas) mushroomed, the value of hydrocarbon reserves skyrocketed and companies that were able to discover and extract oil and gas from the ground became the behemoths of capitalism. It says something that at the beginning of 2016, after BP recorded its biggest loss in 30 years (due to the low oil price, more of which in a moment), Bloomberg reported – with some shock – that the slump had driven the company's market value below $100 billion for the first time since the Gulf of Mexico oil spill in 2010. In IOC terms, 2016 has been a bad year and yet these industrial titans continue to oversee vast global operations and remain amongst the dominant companies in the world.

The rise of the IOC during the 20th century was meteoric. IOCs were, one could argue, the outriders of globalisation (before the term had even been defined). Here were a group of companies who toured the world, competing to buy parcels of land (or sea) – usually from grateful host governments – and who would then take the shortest possible route to producing oil or gas from that land, right up to the point where the economic viability of that operation no longer stacked up. With host governments reaping the financial benefits of selling off – or, in some cases, renting – their land to the highest bidder and IOCs harvesting record profits year on year from the sale of their oil and gas across a rapidly industrialising world, everyone was happy. (Well, perhaps not everyone but more of that later.)

Education and Training for the Oil and Gas Industry. http://dx.doi.org/10.1016/B978-0-12-800980-2.00001-2
1

Within this context of rapid and unalloyed growth, the hydrocarbon business needed one critical thing to keep the oil flowing: people. And it needed a lot of people. From the moment that the first geological surveys were done, through the construction of plants and infrastructure, the operation of wells, the support services across the supply chain, the management and leadership and everything in between, the oil business was an employer – both directly and indirectly – of some considerable heft. And throughout the 20th century, IOCs would approach this challenge – of how to meet the workforce requirements for oil and gas operations – with complete pragmatism. If there were places in the world where they could find, recruit and train people locally – like Texas, Scotland and Norway – then they would happily invest in those local people. But in cases where local skills and talent – and the means to develop and train that local talent – were not readily available, the model was different. IOCs – who commanded a highly trained, mobile workforce across the oil and gas value chain – would import whomever they needed into parts of the world where talent and skills were thin on the ground. They also placed similar expectations on their major contractors and service companies. This approach became the prevailing model for the staffing of oil and gas operations around the world. It had one major downside as far as the IOCs were concerned – the cost of an imported employee or contractor was considerably higher than that of a local, particularly in parts of the world where attracting 'ex-pats' was a challenge (as the 20th century progressed, IOCs gradually threw the net wider in terms of parts of the world where they would set up shop, meaning that, in some cases, operations would be established in areas of political or social unrest or would be in remote areas of the world with tough environments for working). However, multibillion dollar profits can tend to take the edge off a rising wage bill. As the oil – and the money – flowed, there was little incentive for IOCs to change. The short-term approach of the sector allied to a model of operation that continued to deliver growth and revenues meant that few IOCs saw the need to think about doing things differently.

However, this model of operation failed to impress everyone. The link between a country's natural resources and a sense of nationhood and ownership amongst citizens has been well established for over 100 years. One of the first countries to assert this in a meaningful way was Mexico. Following the early interventions of oil companies from across the border in the United States during the 1920s, the Mexican people began to recognise – and then fight for – rights over their own oil reserves. This ultimately led to the nationalisation of the oil industry in Mexico and the establishment of a state monopoly under Pemex. Equally, the experiences of Brazil provide a further window on this phenomenon. The establishment of Petrobras in the early 1950s was partially a result of public discontent with the way in which international companies were dominating the production and distribution of what was very much considered a 'sovereign resource'. This was neatly summed up in the public mantra of the day – 'o petroleo e nosso' or 'the oil is ours'. Although only these are two examples – and, in both cases, examples that culminated in the setting up of national oil companies to oversee oil and gas operations – this demonstrates the historical challenges of the prevailing IOC model.

Essentially, most international companies, faced with a booming and highly profitable industry, saw little commercial value in investing in local people or in building the education and training capacity within host countries to meet the vast and varied skills and workforce requirements generated by their operations. When it came to direct employment, IOCs would much rather fly in ready-made expertise than spend the time, money and effort on developing local staff. The short-term, cyclical nature of employment across the industry – whereby senior staff typically spend 18 months in post within a particular location before moving on – mitigates further against any kind of longer-term investment either in terms of money or thinking. A similar dynamic was at work within the supply chain (where many of the jobs generated by oil and gas operations actually reside). In many cases, the requirements of IOCs in meeting supply chain demands would be met by other international companies rather than by local providers. The rise of the oilfield service companies is testament to this dynamic.

The impact of the 20th century IOC model was clear: whilst it suited the somewhat short-term objectives of oil companies (in that it allowed them to staff operations from an existing pool of trained international staff), this model failed adequately to recognise the role that citizens of a given host nation could and, perhaps, should play in fulfilling oil and gas workforce and supply chain requirements. Furthermore, much of the money and investment that oil and gas operations generated flowed out of the host country with minimal positive impact on the local or national economy. Whilst host governments continued to receive significant payments from licensing rounds and production-sharing agreements, these revenues failed to compensate for the exclusion that many communities felt from what, in large part, was a burgeoning industry. Subsequently, relationships between IOCs, their contractors and local people tended to be strained at best and hostile at worst, and many of the world's poorest had to sit back and watch as global corporations extracted natural resources using highly paid immigrant labour and ignoring the talents and ambitions of the native workforce. In many places, international oil and service companies became unpopular imposters, operating in isolation from the political and societal context within which they were based. For these companies, the costs of operation were high both directly in terms of wages and indirectly in terms of security, reputation and so on. However, if the oil was selling for more than $100 a barrel, this was a price worth paying.

At this point, it's important to recognise the deep sense of awareness that host countries had to this challenge and also important to explore the responses from these countries. It would be insulting to suggest that host governments, faced with an influx of international players into a growing market, failed to understand the wider economic, political and societal benefits that could accrue from effectively managing oil and gas operations at a local level. There are principally two aspects to this response: the first relates to policy and governance and the second to the establishment of national oil companies. We will now briefly consider both.

The case of national oil companies is both complex and illuminating. As a response to the growing monopoly of large international companies, the 20th century saw the emergence of the national oil company, a state-owned and run enterprise that would play a significant role in undertaking exploration and production activities within a hydrocarbon producing nation. For Mexico, this was Pemex; for Brazil, Petrobras. Other parts of the world also adopted this model – Petronas in Malaysia, Saudi Aramco in Saudi Arabia and ADNOC (and other associated companies) in the UAE. The concept of the national oil company was partly born out of a desire for host governments to take control over their own natural resources and to develop a specific department within the government where national experience and skills could be nurtured. From a financial perspective, this would enable a much greater proportion of the wealth that accrued from oil and gas operations to remain in that country. Furthermore, a national oil company would be free to employ whoever it chose in terms of fulfilling direct and indirect employment requirements. As a means of circumventing the dominant IOC model for the industry, one could argue that NOCs had some success, particularly in the second half of the 20th century.

In fact a number of NOCs effectively evolved into de facto IOCs – we think of Petronas and Petrobras, both of which have rapidly expanded (some might say too rapidly in the case of Petrobras) as two apposite examples. However, the growth of NOCs failed to represent a satisfactory new model of operation for the global oil and gas industry. As NOCs became more powerful, we saw the way in which a state-owned monopoly could gradually erode innovation and diminish the ability of a country to capitalise fully on its hydrocarbon assets. Mexico is a case in point here. The inability of an NOC to compete with IOCs on skills, technology or even safety has, in some cases, undermined the NOC model. In Mexico, the notional monopoly of Pemex was, in reality, underpinned by a major outsourcing operation that gave contracts to many of the world's major oilfield service companies. Pemex became inefficient, top heavy and lacking in the kind of professional development that a rapidly changing industry required. We also saw examples where an NOC simply 'played the role' of the IOC, recruiting international talent without really investing in local citizens or in the legal, professional and financial expertise that is the lifeblood of a well-managed hydrocarbon sector. Any new model of operation for the global oil and gas sector should look carefully at the role that NOCs – and, more widely, governments – should play (and we do this later in this book), but we would be foolish to think that we can address the deficiencies of the old model simply by creating a state-owned enterprise and awarding that enterprise the best concessions. History tells us that this does not work.

In terms of policy and governance, awareness of the prevailing IOC model in the latter part of the 20th century – which failed to embrace investment in local people and local businesses – led to the creation of what became known as 'local content policy'. Local content broadly referred to the use of local skills and local companies in meeting the workforce and supply chain needs of the oil and gas sector. Local content policy – usually enacted within a specific law or written into contracts between host governments and operating companies – typically specified targets for local employment and for the use of the local supply chain. In some cases, these policies stipulated the dynamics of ownership in relation to the companies who could bid for contracts across the oil and gas sector of a particular country. In addition, there would often be clauses relating to the investment in local education and training either directly (as in the funding of local colleges or universities) or at an individual level in terms of a commitment to educate or train a set number of employees within a particular company. The Nigerian government were pioneers of local content and set out terms by which international companies should operate within the Petroleum Act of 1969. The act itself had clear ideas about how oil and gas operations should be both staffed by Nigerians and how the companies who won contracts should be owned by Nigerians. However, Nigeria provides us with perhaps the starkest example of how the concept of local content failed to materialise in practice.

The challenges for the Nigerian government, were varied. Principal amongst these was an inability and unwillingness to implement and police the local content regulations set out in the 1969 Act. Targets for local employment remained just that, local ownership of supply chain firms became a way to systematically abuse local content regulation, whereby local businessmen provided a front for international investors. Vested interests in government left international operators with the freedom to operate with impunity. The regulator – such as it was – was powerless and unable to police in any significant way on the rules of operation that it, theoretically, had set. Finally, Nigeria lacked the skilled people in its government and wider public sector to regulate, police and administer contractual agreements between the international oil industry and the government. This lack of human capacity within the government assured capital flight due to the public sector's inability to properly measure local content and thus see where money was flowing. As a consequence, a vast proportion of the wealth generated by the Nigerian oil industry flowed out of the country. And this happened over the course of decades. Whilst the policy itself was all very well, the practical application of the policy was for all intents and purposes nonexistent. Although Nigeria is only one of a number of countries with local content laws – and some have fared better than others – the challenge at the heart of the Nigeria case is a familiar one: essentially, local content laws are a means of a host government trying to impose rules and a model of operation on an industry that has done very well for decades by following an entirely different one. There has subsequently been a long history of international companies finding ways to accommodate (rather than embrace) local content laws and regulations or otherwise circumvent them.

Local content laws have, historically, almost entirely focused on setting out targets for local employment, for local supply chain operations and for local ownership. What these policies typically missed out was recognition of – and a plan to deal with – the realities of the skills, workforce and supply chain base in the host country. Put more simply, it is, one could argue, unreasonable to expect an international company to recruit locally if it turns out that none of the locals are trained and competent to do the job. Furthermore, can we expect international organisations with shareholders to satisfy, and profits to chase, to engage with a local supply chain if there are clear deficiencies in that supply chain or, in some cases, a complete lack of the right type of companies able to meet requirements?

The notion of local employment, local contracting and local ownership is honourable, but the oil and gas business is technical, dangerous and highly skilled. In many parts of the world where local content has been tried as a model, there has been too little invested in the education, training and development of the local workforce and too little focus on supply chain development meaning that the options for international companies in respect of local employment and contracting are, in reality, very limited. Although we can, with some authority, point the finger at international companies for not doing more to support local skills development, we can do the same with host governments.

It is within this context that we believe a new model of operation is needed. The current model no longer works from a financial perspective, and IOCs are beginning to understand the cost benefits associated with preparing local workers for both direct and indirect oil and gas work (more on this later in the book). This model fails to address the long-term needs of any country to develop its people, to build skills, to create education and training capacity locally, to understand the transferability of skills to other industries, to generate employment opportunities where they are needed without that coming at the expense of safety or productivity, to have industries that are enterprising and innovative and that can compete both locally and globally, and to prepare the country either for when the oil and gas or the demand for those resources runs out. What is more, in a low oil price environment that many believe is here to stay, oil companies need to think differently about how they meet their skills and workforce requirements. We believe that the answer to this challenge is localisation. And below, we will explain why.

1.2 WHAT DOES LOCALISATION MEAN IN THE 21ST CENTURY?

Localisation differs from the dominant historical model (outlined in the previous paragraphs) in a number of key ways. Before we explain how, perhaps we should present our working definition of localisation. We define localisation as *the operation of an oil- and gas-producing business which functions sustainably and profitably in the local environment. Its operation contributes to building a local supply chain, maximises the use of local staff resources by strengthening education and training institutions and drives local procurement of goods and services, both technical and operational to the point of maximum long term economic efficiency and sustainability.*

We believe that this model of localisation – which will be presented in more detail in Chapter 2 – provides a compelling alternative business model for the upstream oil and gas industry in the 21st century. This model is compelling – and different from previous models – because.

- **IOCs will be on board** – at the heart of the model is recognition that for a new approach to work, IOCs (and other related industry actors) need to be fully engaged and need to see the direct and indirect economic benefits of adopting and embracing this model. We outline these benefits below.
- **Education and training is part of the solution, not the problem** – a key challenge around localising oil and gas operations is the availability of a skilled workforce to meet the changing needs of the industry and its supply chain. Education and training have often been the missing piece of the puzzle. In our model of localisation, these are the twin pillars on which the model is built.
- **Governments have the know-how and capacity to deliver** – governments work to ensure that local content – that is the value accrued from a burgeoning oil and gas industry – can work to improve the employment prospects of nationals. This means creating the workforce required and creating a strong technical skills base that can transferred into other sectors of the economy and across the wider energy industry. In our model we emphasise the need to share a policy that works, understand how a policy can be implemented given the specific development needs of the country, and developing the policy sharing devices and networks that can greatly assist the creation of value in-country from a growing oil and gas sector.
- **The role of international partners is recognised** – too often in the past, the concepts of local content and nationalisation have revolved around a narrow set of rules that are either grudgingly adhered to or circumvented. In a more evolved model of localisation, the role for international partners – both within the industry and supply chain, and across the education and training sector – is fully recognised and clearly framed.
- **It is long term (because that is the only way it works)** – for a model of localisation to genuinely take hold, it needs time. The short-term approach of both IOCs senior country managers' postings and government electoral tenure has hindered sustained progress towards a more evolved business model across the oil and gas sector. In our proposed model of localisation, the benefits are seen quickly but the real dividend is seen for decades rather than months.

- **There is a legacy that goes beyond oil and gas** – the only model of operation that is sustainable is the one that looks beyond the lifespan of demand for and supply of hydrocarbons and into a postcarbon energy future. For some countries, this future is already close. Our model of localisation recognises the critical nature of this legacy and builds this into the model.

Before we set out in further detail the model of localisation that we are proposing, it may be instructive to clarify why we feel that now is exactly the right time for this model to be embraced across the upstream oil and gas sector globally. Having described the context for the emergence of the IOC model during the second half of the 20th century, we are now seeing an entirely different context within which IOCs, NOCs, oilfield service companies and governments are operating. This context is characterised by a number of critical factors:

- **The price of oil**
  Much has been written and said over the last 18 months about the oil price and the impact that this has had on the sector globally. Much also has been said about the likely price of oil over the coming months and years. What seems clear – when one takes account of global supply, available reserves and projected demand – is that we are now living in an age where an oil price of below $100 per barrel is the 'new normal'. The adverse impacts of the oil price are visible: discretionary spending, particularly in the early stages of exploration, is being cut significantly. Many companies are reevaluating potential projects or are restricting the scope of exploration. Rigs are being taken out of service. In areas where production costs are high, E&P companies are being severely impacted by extremely tight profit margins. In this financial environment, oil companies are keen to reduce spending and bring greater focus to their exploration and production activities.

  However, this strategy – or lack of strategy – does not meet the realities of future energy demand. Most analysis shows that energy demand is likely to increase over the next 20 years. An increasing global population is the key factor behind this. Put in simple terms, this creates a downward pressure on cost. For many fields to be viable, the cost of production needs to be closely controlled and, where possible, costs need to be cut. In the days of plenty, oil companies would think nothing of paying hundreds of thousands of dollars for oil executives and technical staff to be parachuted in to support the development of the latest oil prospect. The incentive for IOCs to recruit, train and employ locally is clear, particularly in parts of the world where local salaries are a fraction of those in places like Europe and North America. As the industry begins to adjust to a new oil price, the compulsion to reduce staff costs will intensify.
- **Political capacity and awareness**
  The world is a very different place today than in the 1960s and 1970s (during which period IOCs evolved into multibillion dollar integrated

companies that turned their attention to the midstream and downstream sectors of the oil and gas value chain). There is far greater awareness of the dynamics of globalisation and the impact that these dynamics have on the lives of citizens around the world. This is particularly true in poorer parts of the world (most notably in Africa) where there has been a steady progress towards greater political stability, a reduction in corruption, increased political capacity and a more progressive and informed approach to economic development. Many of the countries that discovered oil or gas during the last 20 or 30 years are now more fully aware of how and why mismanagement of the industry can lead to the 'resource curse'. The comparative naivety of governments who would happily sell off concessions in return for short-term financial gain now is largely a thing of the past, as international development agencies become increasingly involved in growing the capacity of governments to manage important sectors of their economy. In the 21st century, operators need to recognise the legitimate rights of host nations to benefit from their oil and gas reserves in a meaningful and sustainable way.

- **Social responsibility and ethical business**
  Public awareness of – and public discourse about – the ethics and values that a business exhibits is higher now than at any time in history. Growing public awareness (due to social media and a global media eye) on the energy sector has greatly helped in spreading such awareness. Global businesses now operate within a very different context than was the case in the latter part of the 20th century. It is no longer enough to pay lip service to sustainability and social responsibility through token CSR (corporate social responsibility) initiatives. Companies are judged by how they behave, at home and abroad. And modern technologies have given us the means to police this behaviour. The business model formally pursued by IOCs around the world is no longer perceived as acceptable by the international community, and businesses are now compelled to engage more deeply in activities that contribute positively to the societies where they operate.

- **The revolution in education and training**
  A critical challenge for localisation in the past has been the ability to educate, train and develop local people to fulfil roles across the oil and gas value chain and to ensure that those you recruit locally are able to access ongoing education and training of a sufficient standard throughout their career. Furthermore, a fully localised supply chain can only become a reality if the education and training system is established to help train and develop the skills required to staff that supply chain. Over the last 15 to 20 years, there have been phenomenal shifts in terms of the provision of – and access to – education and training.

  Although there is still some distance left to travel, we have witnessed the 'internationalisation' of education, particularly at tertiary level. This has a number of implications for the oil and gas industry. First, it means that more and more, we can see the quality of available education and training standardising across different territories. As education and training companies and education institutions across the developed world have built their business in emerging economies, the disparity between provision has been highlighted and can therefore be tackled. Second, the influence of technology on learning has been profound with this having particular significance for the oil and gas sector where learning technologies are increasingly used to build technical and vocational skills. Third, the net effect of systemic improvements allied to greater awareness of how we learn has collapsed the time to competency meaning that new recruits can be 'Field-Ready' within a much shorter timeframe. This means that for IOCs looking to develop local skills the pathways to doing this are much clearer than ever before.

- **The shifting geography of exploration and production**
  By some estimates, about 90% of new hydrocarbon production in the next 20 years will come from developing countries (Baker III Institute, 2007). Oil and gas production will shift away from the traditional producing countries and move towards emerging economies. The shift has already begun, with the international oil industry's attention turning to the significant hydrocarbon potential of East Africa – with Kenya, Uganda and Tanzania all vying to become the first producers of gas in the region, and Mozambique not far behind. Given the transformative effects of oil and gas extraction and the opportunities such operations afford for economic growth and national development, governments and development agencies are keen to ensure that local people benefit from new energy sectors in emerging economies. In addition to this, oil companies are increasingly aware of localisation and the possible business impact localisation policies may have on their operations and local subsidiaries.

On the basis of these many a varied factors, we have developed a model of localisation – which we will expand on in the remaining sections of this book – that addresses the failures of previous models of operation and that is aimed at international businesses operating sustainably and profitably within a local context.

The term 'localisation' carries with it certain implications, notions and ideas, depending on the region, area and context surrounding its use. In setting out to create a set of localisation tools to assess the level of localisation, the potential cost-benefits for a given project, and the overall impact on oil and gas operations, it is important to first define what is meant by 'localisation'. The process of localisation involves two elements:

- The first element of localisation relates to local content. Local content refers to the percentage of locally produced materials, goods and services rendered to the hydrocarbon industry. This is generally measured in monetary terms.

Local content, therefore, has a great deal to do with the local supply chain and its capability to deliver goods and services.

- The second element of localisation is related to direct and indirect employment of local people generated through oil and gas operations and activities. Direct employment is the employment with the E&P companies in E&P activities. Indirect employment (which constitutes the majority of job creation) is the number of people employed in the supply chain that operating companies use to procure goods and services.

These two categories offer a concept of localisation and allow for a strategic means of evaluating localisation in a given region, country or for a specific project. Underpinning these two intertwined meanings of the term 'localisation' is the notion of capability, skills, capacity building and workforce development. Immediate localisation is achievable only insofar as the local people, the local supply chain and the local government are ready and able to participate in oil and gas operations. Planned localisation is the extent to which, with the proper metrics, it is possible to localise a business over a defined time period within the metrics of economic sustainability.

Our model of localisation is based on two vertical and two horizontal aspects, highlighted in the diagram below:

GOVERNANCE

WORKFORCE	SUPPLY CHAIN

Education

In Chapter 2 of this book we go into greater detail on each of these elements but, for now, we can summarise each element – and the role that each element plays in our model – as follows:

The first element in our model is 'workforce'. By 'workforce' we are referring to the direct employment that is generated through exploration and production activities across oil and gas projects. This will primarily involve those who are employed by IOCs, national oil companies or oilfield service companies. The range of roles that this term encompasses is broad but covers everything from low- and semiskilled work through technical roles and up to management and leadership. In our model of localisation, the vision for a localised workforce could be defined as follows:

Workforce and skills demands in respect of activities relating to oil and gas operations are met by the local workforce and that this applies across every level and type of job role required in the completion of oil and gas projects.

The recruitment of local citizens into the oil and gas workforce is perhaps the most direct and visible consequence of localisation. By providing jobs to local people, the economic and social benefits are multiplied (more of which below). In exploring this element within any given context (as we do in Section 2.1 of this book) our model considers a series of key questions: what is the availability and job-related competence of the local population? What are the specific (and direct) skills/workforce requirements for a particular operation? What is the gap between the two? What is a vision – realistic and achievable – for localisation in regard of the directly employed workforce? What kind of steps can be taken to achieve the vision of localisation for the directly employed workforce? What will be the benefits to this (economic/social/other)? When applied to a particular context, answering these questions will enable the creation of a strategy for workforce localisation, something we expand on later.

The second element in our model is 'supply chain'. By 'supply chain' we are referring to the goods and services that oil and gas operations typically require as part of their operation. Within this, we exclude any suppliers who work directly on oil and gas operations (for example, oilfield service companies) but include all companies who supply oil and gas companies directly (which may include cleaners, IT services, consulting services, telephony, accounting, transportation, equipment and so on) as well as supply chain requirements that may be generated as a result of oil and gas projects being launched in a particular region (such as construction, engineering, hospitality, et cetera). In our model of localisation, the vision for a localised supply chain could be defined as follows:

The goods and services that are required in order for oil and gas projects to be completed successfully are met by companies that are owned and operated within the host country and, if possible, are located within the region where the oil or gas project is being undertaken.

The localisation of the supply chain is a critical piece of the puzzle in that the employment dividend that oil and gas operations generate is seen most clearly through supply chain demands. Large oil and gas projects generate comparatively few direct jobs once the project has gone through initial development and construction phases. The supply chain, however, is required throughout the entire lifecycle of exploration, production and, ultimately, decommissioning or repurposing. In exploring this element within any given context (as we do in Section 2.2 of this book) our model considers a series of key questions, similar to those explored in relation to workforce: what is the availability, capacity and quality of the local supply chain? What are the specific supply chain requirements for a particular operation? What is the gap between the two? What is a vision – realistic and achievable – for localisation in regard of the local supply chain? What kind of steps can be taken to achieve the vision of localisation for the supply chain? What will be the benefits to this (economic/social/other)?

The third element in our model is 'education and training'. Education and training is a cross-cutting aspect to our model in that it impacts significantly on the two 'vertical' elements previously described. This has two aspects: the first is ensuring that education and training of sufficient relevance and quality is available to local people at every level of operation (and across the supply chain); the second is ensuring that, over time, this education and training is delivered locally, in local institutions with local teachers and trainers and within a system that is able to respond to the changing needs of the industries that it supports. Within this context, the vision for a localised education and training system could be defined as follows:

> *There are suitable, fit-for-purpose education and training opportunities for local people in order to support the goal of developing a workforce that is fully competent and qualified to meet a majority of skills and workforce demands across direct, indirect and supply chain employment. And that these education and training opportunities are delivered locally within a national education and training system that is supported and endorsed by industry.*

As we mentioned earlier in our analysis of the current models of operation, addressing education and training as a core and critical part of localisation has been sorely lacking in the past. In our model of localisation, it is central. Again, this element will be explored through answering a set of questions: what is the availability, capacity and quality of the local education and training offer in relation to requirements for oil and gas operations? Based on the analysis from Verticals 1 and 2, what are the typical requirements for education and training for oil and gas operations? What is the gap between current provision and the perceived requirement? What is a vision – realistic and achievable – for localisation with regard to the education and training provision for oil and gas skills? What kind of steps can be taken to achieve the vision of localisation for the education and training system? What will be the benefits to this (economic/social/other)?

The final element in our model is 'governance and regulation'. This element is similar to 'education and training' in that it is cross-cutting and impacts on all of the other three elements. Governance and regulation explores the role of government ministries and agencies – and the role of the policies and laws that they enact – in supporting (or inadvertently hindering) progress towards localisation. The governance and regulatory environment is critical to any efforts towards localisation. In the vast majority of cases, it is the government that holds the rights to the land (or sea) beneath which hydrocarbons are found. There is a long and complex history of troubled relationships between host governments and international oil and gas companies. Too often, the struggle towards local content has been just that – a struggle between, in one corner, the host government and its citizens' expectations and, in the other corner, the international oil company and its shareholder demands. In our model of localisation, governments are able to develop a clarity of purpose around their own oil and gas sector and are able to articulate that in a way that makes them a partner and friend to the oil and gas industry rather than an adversary. The vision for an effective governance and regulatory environment that supports localisation could be defined as follows:

That the actions, policies and strategies that the government – and its related agencies – follows are coherent in their support for a model of localisation that is long-term and that promotes local participation in, and local ownership of, oil and gas projects and related supply chain activities but that also recognises the need to embrace international partnerships both within the industry itself and within the education and training system that supports and feeds the industry.

The challenge for governments in any hydrocarbon producing country is to balance the need to generate revenues from contracts with international companies with the longer-term goal of supporting national, economic and social development. This is a particularly acute challenge in parts of the world where education and training systems are ill equipped to develop the right level and type of skills required to meet industry demands (the poorest parts of Africa are a good example of this challenge). However, the size, scale and duration of the task should not diminish the will to start the journey. We explore the role of government and regulation in more detail in Section 2.4 of this book and pay particular attention to the role of local content policy, the structure and type of organisations that are needed to support effective education and skills growth, the way in which wider government policy can support localisation goals and the model of contracts that best supports those goals.

1.3 THE BENEFITS OF LOCALISATION

When we talk about 'benefits' within the context of our model of localisation, this means that there are commercial and other benefits to oil and gas companies, financial and developmental benefits to governments and employment and opportunity benefits to local citizens and local businesses. In our model, we believe the benefits to these groups are clear:

Benefits to the industry:
- **Lower staff costs**

  In many cases localising the oil and gas workforce has significant cost benefits to operators, particularly in less-developed parts of the world where wages are typically lower than in the developed world. Analysis of the cost benefits of localising the workforce further demonstrate that even when you factor in the cost of education and training, a largely local workforce is cheaper than an expat workforce.

- **Lower supply chain costs**
  A similar rule applies to the localisation of the supply chain in that engagement with local suppliers (as opposed to international companies) will reduce the costs of operation for IOCs and oilfield service companies. International companies can also maximise their opportunities to secure reliable logistical connections for goods and services through supplier proximity leading to greater project efficiencies overall.

- **Committed workers**
  There is some evidence to suggest that local workers tend to be more committed to their tasks if they are drawn from the location within which oil and gas operations are being undertaken. This is particularly true of parts of the world facing any kind of civil or political unrest. The case of Wintershall in Libya is apposite here as their operations were kept safe and operational by a localised Libyan workforce where other international companies saw their assets deserted.

- **Improved reputation and better relationships locally**
  Oil companies find that they must now demonstrate a 'moral license' to operate. This means addressing the concerns of local people and meeting national expectation around job growth, development and employment. By localising their operations, international companies ensure that they are building positive relationships and improving their reputation amongst the local community.

- **Improved relationships with government**
  Meeting/managing the expectations of host governments in the early stages of an E&P operation is essential. Increasingly governments demand that operating companies demonstrate some level of commitment in the exploration phase. Through understanding localisation, companies can meet government expectations while also creating a foundation for workforce development once a production asset is secured. This in turn works to facilitate the above-mentioned business benefits. Companies can then create competitive differentiation when it comes to bidding and negotiating with host governments and the related authorities.

Benefits to local people:

- **Jobs and opportunities**
  An effective model of localisation will ensure that local people – those living within oil and gas producing countries and within the towns and regions where operations take place – are given the opportunity to become trained and educated so that they are able to take an active role in the

employment opportunities – both direct and indirect – that are created when hydrocarbons are found.

- **Economic development**
  As local jobs are generated – both directly with operating and service companies and within the supply chain – economic development follows. With higher levels of local employment allied to investment in local businesses and opportunities for local entrepreneurs, communities can begin to support wider business growth for goods and services and quality of life for citizens can increase.

- **Sense of ownership and pride**
  The connection between natural resources and the land from which they are mined is long established. Rather than local people feeling excluded and, therefore, resentful of international companies, they can feel involved, engaged and enthusiastic if they are given and can take the opportunities to participate in a meaningful way.

Benefits to the country

- **Keep the money in the country**
  An effective localisation model – which embraces international operators but is underpinned by partnership and collaboration on a local level – can ensure that the economic benefits of the industry as a whole can be felt more widely. Just as the local economic development mentioned previously will be felt within those communities, a greater participation by local citizens and local companies will drive up tax revenues and GDP.

- **Skills legacy**
  By ensuring that there are local people who are directly and indirectly employed in oil, gas and related operations, that population will develop knowledge, skills, experience and expertise that will outlast the oil that is in the ground and that will be vital to the diversification of the economy (a subject we expand upon in the third section of this book).

- **Education and training legacy**
  In addition to a clear skills legacy, the model of localisation that we are proposing will also ensure that the majority of education and training received by those working across the industry is delivered in the host country. In some cases, this will involve significant investments across the whole value chain – from facilities to teachers and training programmes to accreditation. The value of this – above and beyond the capacity to educate and train for oil and gas operations – is that this capacity will be transferable to other economic and industrial areas. Once you have a strengthened education and training system, you can build from there. This education and training legacy is one of the most compelling reasons to invest in our model of localisation.

1.4 THE CHALLENGES OF IMPLEMENTING LOCALISATION

Despite our passion for the model of localisation that we are proposing, we are aware that there are very good reasons why such a model has, so far, failed to become established across the industry (although, we would suggest, we are beginning to see these ideas take root). In recognising some of the principal barriers to localisation, we must be open and discover solutions that really work in addressing these barriers and not be led by dogma.

You will notice in the list below that some of the barriers we identify are a mirror of some of the factors we highlighted as being supportive of the localisation model:

- **The cost of oil**
  Although the cost of oil is identified above as a driver towards developing a new model of operation, it is also a hindrance. The prevailing mood of the industry within a low oil price environment is one of caution. Furthermore, education and training budgets are generally one of the first to be cut. Part of our model of localisation has to address this head-on and prove to reluctant operators that an investment in local talent and the local supply chain will be part of a cost reduction strategy and one that will reap rewards quickly.
- **Commercial expediency**
  Any investment in education and training – and more broadly in exploring new models of operation – will invariably involve time, effort and patience. These are all in short supply where prices are low and the need to generate immediate revenues and profits is pressing. Embracing localisation will involve something of a medium-term view and, in the current climate, many may feel they do not have the luxury of looking beyond the short-term.
- **Companies do not want to change**
  The instinct to 'carry on as normal' has, in the past, been driven by the fact that the previous model was considered to be broadly sustainable and revenues were high. However, even in an evolving market where the need to adapt is compelling companies can be very resistant to change. This is often an issue of culture and, within that, a challenge of leadership. There have been some indications that both culture and leadership across the industry are changing and this may see a greater sense of optimism that progressive models of localisation may gain more traction.
- **The shifting dynamics of operation**
  We are beginning to see a rise in demand for more technically complex skills and expertise, particularly in relation to the extraction of unconventional energy sources. Even though the cost of unconventional production tends to be higher than for conventional hydrocarbons, there is still significant investment going into unconventional energy projects globally. The downside

for localisation is that those involved may see the technical requirements for these projects as being even further beyond the talents and abilities of local people. However, a clear assessment needs to be made of the direct technical job roles that these activities require set against the much higher demand for support workers and supply chain staff which would remain largely unchanged from conventional energy projects. There will always be a need to import high-level and scarce expertise when needed from wherever it is best found.

- **Operating companies are businesses**
  Our model of localisation involves, at its heart, recognition from both international operators and host governments that education and training has to be a central point of collaboration between all concerned. It cannot be left up to the local system and the local government to take care of the skills development challenge. Although the mood is, perhaps, changing, operating companies have often met that request with the refrain that 'we are oil companies, not education companies'. Again, the challenge here is to overcome cultural problems within the business and to ensure that the leadership is on board.

- **Political capacity of some host governments**
  Our model of localisation very much aims to support the efforts of host governments in creating the right context for local participation in the oil and gas sector. However, we need to recognise that the role of government – which we look at in detail in Section 2.4 of this book – is critical and that the history of international oil and gas operations tells us that the capacity of host governments to be effective regulators, educators and managers of and for the industry is patchy. Part of the challenge in making our model of localisation work is to build that capacity and to support governments in developing their own competencies in order to deal with the issues and demands that localisation generates.

- **Education is a long road**
  For all concerned, there has to be recognition that the job of building a local workforce and supply chain that is capable of meeting the demands of oil and gas production in the 21st century is a long project. Education takes time and the results may not be visible at first. This is where leadership at both a political and industry level needs to be strong. Long-term plans that start delivering in the short-term are key.

With these challenges firmly in mind, we recognise the task that lies before us: what we must do is create a mechanism by which we can measure, understand, quantify and analyse the factors that affect or that define localisation and we must present this in a way that everyone can understand. In doing this, we can then help governments, operating companies and others to develop a coherent strategy – one that is defined at a local level – to ensure that all key partners are able to work towards a desirable and achievable vision for localisation based around clear objectives that serve the industry, the government and the citizens of the host country. In the next section we set out a detailed plan for that mechanism and explain, with the help of some specific case studies, how we think that mechanism can be implemented in oil and gas producing countries across the world today.

Chapter 2

Understanding Localisation

Chapter Outline

Education and Training for the Oil and Gas Industry. http://dx.doi.org/10.1016/B978-0-12-800980-2.00002-4

2.1 INTRODUCTION

In Chapter 1, we explored our rationale for localisation. The challenge we have now is to explain our model of localisation. In this section, we will provide a more detailed definition of what we mean by 'localisation' and we will follow that with an analysis of the four key elements that impact on localisation – workforce, supply chain, education and training, and governance and regulation. At this point, we should emphasise that, like any business model, our model of localisation is constantly under development and review. We have invested significantly to develop the thinking that you see across these pages. We also recognise that there is more work to be done.

2.1.1 Understanding 'Localisation'

The term 'localisation' carries with it certain implications, notions and ideas, depending on the region, area and context surrounding its use. In setting out to create a model of localisation that assesses the level of localisation, the potential cost–benefits in a given project and the overall impact on an oil and gas operation, it is important to first settle on a working definition of 'localisation'.

Localisation can fundamentally be understood in two ways:

- Workforce localisation: the direct employment that is required by international oil companies (IOCs), national oil companies (NOCs) and the oil field service (OFS) industry to physically explore, extract and produce oil and gas.
- Supply chain localisation: any goods and services required by IOCs, NOCs and the OFS industry to support their operations in country.

In the first instance the emphasis is on the direct workforce and the localisation of employment opportunities within the sector. This pertains to how IOCs, NOCS and international operators go about recruiting and developing their local staff on a given project or around a specific asset. Mainly, direct employment is calculated in terms of a company's direct workforce and all people employed by the IOC or its local subsidiary. However, for our purposes, we have shifted the perspective slightly to include all the employees hired by IOCs including those employed by OFS companies in a given country, project or asset. These workers, in the case of the OFS industry, encompass all those required to undertake essential, on-site tasks including those within technical or evaluative roles. For the IOC's, however, the direct workforce relates to the entire local workforce employed to run the business from either a technical perspective or a business and administrative perspective.

In thinking about supply chain localisation, the emphasis is on 'localisation of procurement'. IOCs, instead of drawing on international vendors to procure goods and services integral to their business operations, look in-country to find the goods and services they need. Supply chain localisation has wider implications for the national economy and is therefore considered to have a larger impact on the host nation's development. The localisation of procurement is often framed within 'local content policy' as relating to the percentage of locally produced materials, goods and services rendered to the oil and gas industry. This is generally measured in monetary terms. Local content, therefore, has a great deal to do with the local supply chain and, ergo, depends on the supply chain's capacity and capability to deliver goods and services to a standard and at a scale that meets industry needs.

These two elements, though not entirely separate, provide a broad definition of localisation and therefore give us a strategic means of evaluating localisation in a given region, country or for a specific project. Underpinning these two intertwined meanings of the term 'localisation' is the notion of capability, skills, capacity building and workforce development. Localisation is achievable only insofar as the local people, the local supply chain and the local government are ready and able to participate in oil and gas operations.

Localisation policies – sometimes defined as 'local content policies' – can and do differ significantly in their structure. Some are more prescriptive while others work as guides alongside individual agreements with IOCs whereby specific localisation targets are stipulated within those contracts. However these policies and agreements differ, their basic aim is to secure economic development through capitalising on oil and gas operations and creating 'local content' which, measured via various metrics related to direct workforce and the supply chain, fundamentally means creating and retaining value in the host nation. Capital flight is generally a consequence of a shortage of inward investment into local workforce capacity building (or workforce development) combined with international companies' sidestepping regulations around local participation. This typically occurs because the local content policy is either too vague or too prescriptive, because host nation governments fail to be transparent in their dealings with the oil and gas sector and its national/international participants, and because attempts to create value by investing in the education sector fall short [particularly in respect of the higher education sector and the technical and vocational education and training (TVET) sector].

We have consistently seen how local content (or value added to the economy through natural resource production) can only accrue if a genuine and sustainable model of localisation is pursued. This might best be represented in the following way:

Local content = national workforce development + revenues from natural resource production

This relationship must be understood from the outset, as it is the premise that underpins why localisation – and the need for a coherent localisation strategy – is so crucial to governments of oil- and gas-producing countries. Localisation, at its simplest, means local people are able to work directly or indirectly in the oil and gas industry. This requires those people to be fully equipped with the knowledge, skills and behaviours needed to enable the industry to operate safely and productively. In many parts of the world, this is not currently the reality. The main problem undermining effective localisation is a dearth of local talent and the absence of a national workforce capable of undertaking work across the industry. In the supply chain area, it may be that the country's capacity to manufacture and fabricate goods is limited, unfit for purpose or suffers from subpar quality. It also might be that essential goods and services that the oil and gas industry relies on in the wider economy are simply not available locally as the oil and gas industry may be comparatively new and thus there has not previously been a market for certain goods and services. In some cases – Mozambique springs to mind – the lack of economic and industrial development across the country means that the local sourcing of goods and services is virtually impossible. The need to seek out international vendors and rely upon a workforce that is largely imported from overseas raises the costs of doing business for IOCs, who then struggle to meet production targets when the commodity price dips. This then threatens the viability of projects and can cause international companies to cancel investments. In turn, economic development is hampered by slower-than-expected oil industry growth rates, and the country struggles to meet development targets due to capital flight.

However, the ability to create and retain value in country largely depends on how 'field-ready' the local workforce is for oil and gas operations, and this requires concerted efforts from a policy perspective to support the development of oil and gas competencies amongst local citizens directly as well as strengthening education systems so that they can produce competent nationals with the right knowledge and skill sets to be employed across the industry.

Who Is a National?

How to define a 'national' is a key question when it comes to measuring the impact of localisation strategies. Indeed, our terminology could cause confusion for some as what we are proposing is closely aligned to the concept of workforce 'nationalisation' (a concept itself that is often articulated by the addition of the suffix '-isation' to the nation in question, for example, 'Saudisation' or 'Emiratisation'). Usually a national is defined as an individual with rights of citizenship, and – for the purposes of this book – one who is on the company payroll and who is resident in the country. For international oil and gas and OFS companies, this definition may not fully capture the extent of local employment, since some of their 'national' staff may be working on assignments abroad. If a multinational oil company has a local workforce in, say Ghana, and has spent time training particular skill sets and the IOC in question takes a number of these employees to fulfil operations in another part of the world, can they still be counted amongst the local workforce of Ghana? Certainly, our focus is to ensure that oil and gas activities generate real benefits for citizens in host countries. If that means that those who are trained locally end up working internationally, then we would consider that a success. However, for the purposes of understanding the extent of localisation within a particular location, we will consider only those who are actually working in the country of origin.

Localisation can run into obstacles around how a 'national' is defined for other reasons. In any given country, regional demographics, cultural associations and religious or political affiliations may complicate a localisation strategy. International operators may believe their localisation strategy is working because of the numbers of nationals being employed and trained. However, if these nationals come mainly from one group of people, another group may perceive this as a form of favouritism (or even prejudice) which may fuel existing sectarianism. This could be as simple as Sunni–Shia divisions in Iraq. However, these divisions can create serious issues for IOCs. While this book considers a 'national' as anyone resident within, and with citizenship of, a particular country – and while we avoid addressing the issues of cultural divisions that may exist within a country – it helps to be aware of this complexity. It should be understood that, for the purposes of our model, localisation looks at nationals and does not differentiate between particular ethnic groups or cultural associations. For example, in Iraq we would simply look at the number of Iraqi nationals employed by the sector as opposed to differentiating between Kurds and Arabs or Shiites and Sunnis.

Similar problems can arise when defining what we mean by the term 'employee'. In measuring the local employment levels within an IOC, we may neglect to account for locals who are employed by an outside contractor (such as a recruitment agent) and brought in to work for the IOC. This is especially so when wages are paid by the agent and not the IOC.

2.1.2 The Life cycle of Localisation

Localisation is an integral part of project management. But this involves an integrated and holistic understanding of a project's life cycle and should be part of a strategic approach to workforce development that can – and should – accrue financial benefits to the operating company. This will create significant benefits in terms of economies of scale and through building cohesive collaboration with local service providers. Let us say a typical operator producing 100,000 barrels per day requires 100 technical staff. As the resource base matures and evolves, the focus of expertise for those staff will change.

But developing local people for direct and indirect (supply chain) roles is not a short-term endeavour. It requires long-term strategies that fit into the overall project life cycle. Statoil needed 14 years to acquire the skills to become the major operator it is today. During that period, it hired 8000 staff and took 8 years to turn a profit. Ergo, localisation must be viewed as a long-term endeavour that seeks to address the various phases in a project (or asset's) life cycle. With this in mind, it is worth briefly discussing the various stages of a project life cycle:

- Stage 1: The 'Exploration and Production' (E&P) stage, looking specifically at the early stages leading up to and beginning to develop a production asset.
- Stage 2: The 'Increasing Production' stage that looks at the mid-stage of a project life cycle where increasing production is high priority.
- Stage 3: The final stage, the 'Maturing Asset' stage, where the nature of the asset is shifting and, along with this, so too does the nature of skills demands change for the project. Ultimately Stage 3 ends with the decommissioning of the asset, which is also a skill-driven part of the project.

Stage 1: Exploration and Asset Establishment

In the predrill planning stage, our model of localisation focusses on enabling operators to access local well technicians and well engineers to meet key subsurface objectives. In addition to direct employment of local people/local suppliers, operators need to assess strategies relating to their commitment to localisation and they will need to develop a continuing education and training strategy that will meet government expectations while ensuring that there is a direct benefit to the pre- and postdrilling phases.

In the construction phase, the opportunity to build on existing skills is multifaceted. However, utilising the existing skill base and workforce requires strategic alignment with the oil and gas industry and necessitates the proper distribution of highly skilled people.

Stage 2: Production Stage

In the production phase the emphasis is on maintaining and raising the production of well sites as well as meeting regulatory requirements around environmental safety. In addition to this, other skill requirements come into play – asset integrity, data collection and reporting, IT requirements, legal and professional services and so on.

Stage 3: A Maturing Asset

In the final major stage of a project, when the asset is maturing, companies need to invest more in the project. Usually the skills needs shift, with importance being placed on maintaining the integrity of the asset while bolstering and promoting production. For example, in the case of an ageing oil field, trying to maintain, and even heighten, levels of production usually involve employing one or several enhanced oil recovery (EOR) techniques. EOR techniques, if not properly administered by those with the competence to plan, execute and supervise a project, can significantly damage the asset, compromise the safety of the on-site crew and shorten the production life span of the asset in its remaining years.

Due to the nature of oil and gas operations, the asset can become more dangerous to operate. This can be a consequence of ageing infrastructure and outdated rigs and platforms that are reaching the end of their operational life and present more dangers. This trend, whereby assets go from being highly dangerous in the first stage, then shifting to high levels of productivity and low levels of danger, and then – in the final stages – return to high levels of danger and low levels of production, is known as the 'bathtub curve'.

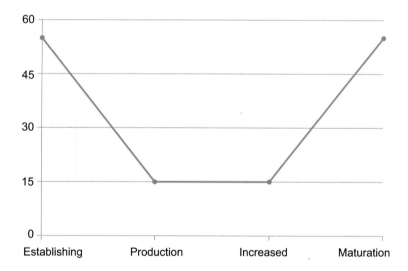

The bathtub curve means that oil companies generally run more risks at the beginning and end of a project's life cycle. Complex procedures and operations bring with them inherent dangers, as does ageing equipment and infrastructure. Of course, this means that the requirement for skills in relation to areas like well maintenance, asset integrity and EOR techniques must be planned for early on to facilitate the safe operation of the asset in its later phases. The bathtub curve allows us to see where the real demands and challenges are around direct workforce development for oil and gas operations.

2.2 THE KEY COMPONENTS OF LOCALISATION

In this section we will explore the key components of localisation, examining what areas oil and gas companies and host nation governments must analyse if sustainable employment and value retention is going to occur.

An abundance of oil and gas – or indeed any natural resource – does not naturally transform a developing economy into an economic powerhouse. In fact, throughout history we have seen more examples of the 'resource curse' than we have seen meaningful economic transformation. This, from the perspective of government, has tended to point to a failure to understand how people, and the skills they have, are central to economic development. Although this can be – to some degree – attributed to a failure to fully understand the dynamics of the industry, poor handling of the economy and toothless local content policies, some of the responsibility lies with oil and gas companies who have preferred to draw on their existing talent pool instead of investing in local people and building a skill base near to their operations.

In today's E&P business – where we commonly see a reliance on expat labour coming from developed (usually North American/European) countries – oil and gas companies are struggling to lower production costs and protect profit margins. The principal way for IOCs to cut costs has been through reducing OFS contracts. This has resulted in shrinking workforces and a reduction in the commissioning of important engineering services with companies like Schlumberger, Baker Hughes and Halliburton suffering as a consequence. Localisation plays an important role here. If IOCs can reduce their reliance on expats and build local expertise, they can improve their workforce, reduce costs, diversify their labour and create value retention for the local economy. This then feeds a growing supply chain for the IOCs to draw on for goods and services, which in turn lowers costs and diversifies the IOCs' supply chain. This cycle of investment and reinvestment benefits both the host nation and the IOCs who are – particularly in a low oil price environment – seeking to reduce their reliance on large international vendors who can provide complete, though expensive, full service packages.

Awareness of these new models of operation have been most acute in the new oil and gas nations of Africa with host governments and international development agencies demonstrating a clear understanding of the significant development opportunities that can accrue. Over recent years we have seen an evolution in the way policymakers approach local content regulation, with the aim being to utilise oil and gas resources, which are inherently limited, to fuel wider economic development. Local content policies are now a political priority in many resource-rich developing countries, regardless of whether those countries are mature producers or recent entrants into the global oil and gas market. The application of local content policies raises a number of interesting questions:

- What is 'local content'?
- How can 'local content' be measured?
- How successful are local content policies?

- Is local content useful?
- How is local content applied from a practical perspective?
- How does the oil and gas sector react to such policies?
- What are the costs and benefits of introducing local content policies?

Is it instructive to address some of these questions prior to examining the components of our localisation strategy (which we present as a distinct and compelling alternative to what is currently the norm). When discussing local content policies, it helps to have a working definition of what they are and what they are composed of:

- Local content policies have two coinciding concerns: first, they are concerned with the initial immediate increase in the percentage of local employment. To facilitate these immediate goals, IOCs often hire low-skilled workers (such as drivers and transporters) into their direct workforce without considering an investment in skills further up the pay scale. Second, local content policies are concerned with actions that will lead to longer-term increase in local participation (such as workforce development, the provision of training to nationals and skills building). Local content policies themselves do not lead to increased local content, but if actualised they can create a skilled workforce with the right capabilities to undertake a range of roles across the energy sector, as well as enhance the development of transferable skills. Cross-sector training and transferable skills are highly important, and local content policies can also support the development of other industries that have a 'natural synergy' with the oil and gas sector.
- Understanding the term 'local' is hugely important. When determining national, this is relatively simple and straightforward once an agreed definition is in place (see previous). However, local content policies usually make some reference to the use of 'local' companies. This is open to a range of definitions. Companies in the supply chain who are engaged in the supply of goods and services may be locally based and owned, locally based but foreign owned (in total or in part) or locally owned but located overseas or abroad. Understanding these distinctions is necessary to a study and evaluation of local content and in-country value generation and retention. The direct benefits each of these brings to the national/local economy differ. As an example of this, foreign investors may be unwilling to invest in companies that are primarily locally owned. This can even hinder an investor's willingness to facilitate technology and knowledge transfer. For some purposes domestic ownership can work against attracting foreign investment. For these reasons, when looking at local content and the supply chain, it will be necessary to take into consideration the ownership of the company and qualify it as truly 'local'. A deeper discussion of this will take place in Section 2.2.2.4 (Classifying local suppliers), and the case study on Nigeria will serve to highlight some of the dangers that a host nation may encounter if company ownership issues are ignored.

- Local content – as noted by the World Bank and other development agencies working in this area – can refer to value added or jobs created anywhere in the economy as a result of an action taken by an oil company, or it can refer to something that takes place more immediately, say, in the local vicinity of the oil production plant or refinery. While local content policies generally do not differentiate between these two areas (and for the purposes of our model we will take into consideration the national economy, not limiting ourselves to the locations of production), it stands to reason that those closest to major oil or gas production facilities will exert the most pressure to find employment in the sector or see benefits from production. In Iraq, this has certainly been the case, where disenfranchised local communities will protest at oil production sites, often causing delays and halting operations.
- Local content must also be considered from the purely corporate social responsibility (CSR) point of view. Donating funds for hospitals, medical equipment and education infrastructure, all help to add to the local economy and its productivity although the connection between the concept of genuine local investment with CSR is, in many ways, highly suspect and generally works at a level that has very little real impact on development or on local participation.
- Local content must also take into consideration the ongoing movement towards fully integrated oil and gas businesses, where oil companies no longer just pump oil and gas out of the ground, but build and install refineries to create value-added consumable goods, such as fuels and lubricants. These can greatly benefit the local community, providing cheap, easily accessible energy. Understanding the backward and forward links (the principal backward link of a refinery is oil production, and the principal forward link of oil production would be a refinery) is necessary. However, for the purposes of brevity in the book we focus our discussion of local content on the perspective of the upstream segment of the industry.
- Local content will vary over the life cycle of an oil or gas project and across the various assets held by companies. Going through phases of exploration, construction, production and maturation, the requirements both in terms of direct workers and the procurement of essential goods and services will change and fluctuate. This naturally has an effect on local content, causing it to vary over time. Any comprehensive approach to managing and developing a localisation strategy must take into consideration these changes and build them into the strategy.
- Likewise, any local content policy must take into consideration where the project is. An example of this might be in the early stages of exploration when a new discovery is made. Due to geological uncertainties, the company and the host government may not be certain what quantity of

hydrocarbons exist, and therefore the company will understandably be unwilling to invest in local content and localisation. Where the rationale is weak for localisation, oil companies will not invest in the local supply chain or their direct workforce locally.

The Difference Between Local Content and Localisation

So far we have used two terms: localisation and local content. At this stage it is now important to understand the difference between these two important terms to frame the wider discussion of how we approach localisation and building local content.

Local content and localisation, though they share many of the same aspects – particularly with regard to skill development and workforce capacity building – are not interdependent. They do, however, share a symbiotic nature in one true sense: localisation cannot be achieved without, even unintentionally, creating local content. However, local content and localisation represent two different end goals.

Local content: Taking into consideration the above points related to local content policy, it would be accurate to suggest that local content is about meeting the national development objectives of the host nation through specifying certain levels of job creation, employment and the use of local companies. Here the emphasis is on developing the capacity of the host nation to run the oil and gas sector and supply it with goods and services. The objective of this is to create value in country, increase economic growth and prevent capital flight. There are tangible economic and monetary measurements that reveal the extent to which local content is being achieved or built.

Localisation: Localisation, though it entails some of the same activities that lie behind local content (such as workforce development, employment, local procurement, supply chain capacity development, etc.) is not limited to the singular goal of creating value or adding value to the host nation. A major issue for local content policy is that it fails to recognise the commercial dynamics of the oil and gas sector. Our model of localisation is about achieving the business goals of the IOCs or international companies investing in the country alongside an achievement of national development goals. There are clear objectives for any business, and meeting the need to lower operational costs at the site as well as finding the right sort of expertise is now a critical consideration for the global oil and gas industry. Reducing costs can only be sustainable and safely achieved if localisation – to the extent to which makes sense for the business, region and operational circumstance – is strategically driven and at the core of the upstream sector.

To summarise, local content is about creating and adding value in country through engaging locals in both the direct energy sector workforce, while localisation meets that objective whilst also creating value for the company and/or its local subsidiary through the employment of nationals in its direct workforce and engaging local vendors in the supply chain.

2.3 THE 'WORKFORCE' COMPONENT

As discussed previously, one of the key components of localisation is the direct workforce employed by IOCs and international operators coming into the host nation. In this section we will discuss the following:

- We will focus on those who are directly employed in the delivery of oil and gas operations, either in terms of those employed by IOCs/NOCs or those employed by OFS companies.
- This will be the distinction between 'workforce' and 'supply chain'.
- As outlined in the introduction, we will look at the following in relation to workforce:
  - What is the current model/approach to fulfilling direct workforce and skills demands for the oil and gas industry globally?
  - What are the failures of this approach, and how are these limiting oil and gas operations and the economic development of host nations today?
  - What would a model/approach driven by localisation look like?
  - How do we understand/quantify/measure localisation in terms of direct employment in oil and gas operations?
  - What would be the benefits to adopting a localisation approach is regard of workforce?
  - What would be the challenges/barriers to adopting a localisation approach?

2.3.1 Workforce: The Current Model

Direct employment in the oil and gas industry does not, at the moment, create the level of employment required to support significant economic development in emerging economies. While the emergence of an oil and gas sector – as we are now seeing in parts of East Africa – may generate a several thousand new jobs, not all of these will go to locals. The international oil and gas industry still relies on an international workforce for scientific and technical experts. Additionally, in the very early stages of an E&P project (where an IOC may have limited rights to explore for commercial oil or gas across specific lots of a known reserve), there will be minimal investment in local talent from the IOCs involved at this point.

Generally, taking into consideration the project life cycle from exploration to asset establishment and through to maturation of the asset, the number of workers required fluctuates, as do the specialisations and skills requirements. According to data we have collected – and interviews with members of the Getenergy Upstream Group – we have judged a typical oil and gas operation to have the following people requirements across the life cycle of an operation, taken into four broad categories.

With this in mind, the current model typically looks like this:

- **Exploration stage**: In the exploration stage, direct workforce requirements are very small. Generally, IOCs rely on OFS companies to undertake drilling and other essential core operations as they go about searching for commercial quantities of oil or gas. Consequently, there is typically very little – if any – investment in building a direct workforce for this phase, since there is not an asset to establish and produce from. IOCs prefer to rely on international OFS companies to undertake work in this stage – or on their own experienced staff or a combination of the two – and the opportunities for localisation (at least at the technical and operational level) are highly limited. At this stage, however, localisation can work in the IOC's favour in other areas. Designing contracts requires lawyers, translating contracts requires translators and negotiations may also require the use of skilled interpreters. In addition, IOCs establishing a presence in the country may also tap into the local talent pool to expand their administrative and professional staff. Identifying drilling sites also requires highly skilled scientists and professionals who are able to interpret geological data to plan their E&P strategy. This also demands an understanding of the testing process and utilising low-cost measures to gain the most accurate data, such as soil gas surveys. At this stage, IOCs will also draw on their own pool of talent to produce drilling proposals that will include key dates for asset establishment and identify potential issues with drilling, environmental dangers/hazards and anything that may hinder oil production. It is at this point that IOCs could be evaluating and designing their localisation strategy.

When digging exploratory wells, skilled survey crews will be required, as well as drillers. In an offshore play, this may also involve the use of other teams of skilled professionals, particularly seasoned maritime specialists and electrical crews. In today's upstream business model, oil companies are increasingly reliant on OFS companies to provide the expertise to carry out exploratory drilling projects. OFS companies have been providing engineering services to the major oil companies (both IOCs and now even to NOCs) since the 1980s.

- **Asset-establishment stage**: Once commercial quantities of oil or gas are found, the use of technically skilled and competent people is required to establish the asset. This means installing rigs, which can vary in complexity depending on the location and geological circumstances. It means establishing a functional well that can be safely produced from and meets international environmental standards. When establishing an asset, IOCs will also have to consider the extent of existing infrastructure. Producing oil or gas is only step one in a much larger process, where the hydrocarbon must be stored, transferred and transported. Without the infrastructure in place, IOCs need to consider how this infrastructure can be developed.

Throughout this stage and into the next stage, which focusses on production increase, the following steps are necessary to produce oil or gas from a well:

- 10–30 different service companies are required to build a rig, drill, and maintain the well. Each company working on a well must adhere to around-the-clock scheduling, safety and environmental practices.
- Developing infrastructure required by the asset, such as building roads or constructing buildings, fences, ports, etc. will also need to be addressed.

This can vary greatly depending on where the asset is located. For example, there will be huge differences in the challenges relating to developing infrastructure around an offshore asset and an onshore asset. Water and electricity infrastructure will be part of the challenge.

- Part of the skills and workforce challenges will also need to focus on clearing the area for the new rig. This can mean taking measures to prevent environmental damage. Onshore, this could mean digging pits to prevent water table contamination.
- Drilling and digging may not always be focussed on readying the well for production. Such activities could extend to holes required to stabilise equipment and pipe during the drilling.
- A rig that can dig a 10,000-ft well requires 50–75 people and 35–45 semitrucks to move and assemble the rig. In the current oil and gas upstream model these people are generally provided by OFS companies, who will have a ready-to-deploy international team already assigned to the task. The project, at all stages, will be overseen by representatives of the IOC (or the NOC, depending on who contracted the OFS company).

- **Production stage**: In this stage the focus is on improving the levels of production from an operational asset. At this stage, IOCs predominately rely on OFS companies to fulfil the bulk of their manpower requirements in terms of engineering services. However, OFS companies are required to meet specific localisation targets, which are usually set by the IOC (as a consequence of the local content policy that they are required to adhere to). At this stage less specialists are needed, since production is well-established. However, low to medium skilled individuals are required to meet the needs of ongoing maintenance, and competent technical people are required in leadership and management roles to oversee operations (such as the installation manager, or IM, who is typically employed by the IOC).
- **Maturation stage**: In the maturation stage the need for specialists rises. The two most technically complex and demanding phases of an oil and gas project are the beginning (establish the asset) and the end, when the asset enters a period of ongoing maturation. In the maturation stage, producing from the field brings with it more risks. Additionally the skills required to manage these risks while maintaining commercial levels of oil and/or gas production require more specialised workers. At present, this stage is generally undertaken again by suitably qualified OFS companies.

2.3.2 Workforce: The Failures of the Current Model

The current model of direct employment in the oil and gas industry is one that relies predominantly on international workers being moved to new areas of operation. Oil field service companies often supply the majority of workers, particularly drillers and technical specialists in areas such as seismic, characterisation, completions, subsea production, well intervention and so on. However, IOCs will usually have a small number of their own senior management and project support staff who will help to supervise the project or manage the asset. These people include the installation manager and project managers, though OFS companies are increasingly taking responsibility for overseeing project management and design.

In this model there is a clear mindset, which has dominated the oil industry since the 1970s and 1980s. The idea is to contract companies with the technical skills and technology to quickly move a hydrocarbon discovery into a production asset for the IOC (or even an NOC), and then manage this asset so that it produces where the seismic data indicate the reserve is capable of. In today's industry, there is one important addition to this approach: an emphasis on safety. IOCs often justify large contracts to OFS companies by citing the inexperience of local companies and their inability to operate safely. OFS companies, however, are still dependent on partners to assist in noncore engineering tasks. Schlumberger, as an example, works with transportation partners to transport equipment and waste. The current model of workforce employment in the oil and gas upstream industry (in a developing economy) can be broken down in the following way:

In the earlier description we can see that the level of involvement of local workers is negligible when compared to the number of international and regional workers sourced through OFS companies who typically have large numbers of workers readily available to undertake specific tasks. It should be noted that OFS companies are increasingly responsible for providing the technical personnel required. This presents significant cost issues, which are highlighted when oil and gas prices plummet (as happened in 2014).

The IOCs have begun to move away from conventional oil production and now around half of IOC spending goes on unconventional or deepwater oilfield development. This is primarily due to IOCs finding it increasingly difficult to strike production sharing agreements (PSAs) and licences with NOCs, who have become bigger and stronger and now dominate some of the world's most productive fields. Some NOCs still struggle with their assets and operations and then enter into agreements with IOCs, but many have learnt how to produce by themselves. Successful NOCs are now able to produce oil and gas without help and around the world. The governments that own and manage these NOCs can use their commercial success to build cash reserves that can then fund further oil and gas operations alongside supporting workforce development. However, large NOCs are increasingly turning towards international OFS companies. In Saudi Arabia, Saudi Aramco turned to OFS companies due to the fact a small proportion of nationals are involved in the industry at the technical level. Consequently, Saudi Aramco depends on an almost symbiotic relationship with Hercules, a major OFS company, to provide the majority of its manpower needs in-country and to help develop Saudi Arabia's offshore interests.

Saudi Arabia

Saudi Aramco is one of the biggest oil companies in the world. While its contribution to the Saudi economy is significant and the country has a strong demand for workers, only 20% of the workforce is local, with at least 60% being nonnationals. To combat the potentially disastrous consequences of a low oil price, Saudi Arabia in 2016 announced the Vision 2030 plan, which seeks to reduce the Kingdom's reliance on oil rents. One of the major criticisms at the time of writing was that the Vision 2030 plan failed to address the key concern of attracting more locals into the technical roles that the country's significant oil and gas sector relies on.

However, Saudi's leaders are aware of the need for diversification, and the country is seeking major investments from leading energy companies into key segments of its economy to develop the skills and technologies needed to carry the crude-rich country into the future. These have included significant investments from GE (General Electric) in technical and renewable energy areas.

In recent years, the Kingdom has set about a project of 'Saudisation' to increase the number of nationals working in key areas across the most important segments of the Kingdom's economy. This process has had to tackle challenges around local participation that have traditionally limited the number of Saudi's taking up technical roles in the hydrocarbon industry. To do this, the Kingdom has increased investments in technical and engineering subjects, which is hoped will help to shift the somewhat negative perceptions of a career in oil and gas. To date the Kingdom has invested in the creation of the Saudi Petroleum Services Polytechnic (SPSP), the National Industrial Training Institute (NITI) and the Colleges of Excellence (though not directly related to the oil and gas industry, these colleges do serve as technical centres which will help to ready people for work in the wider oil and gas supply chain).

Estimates have shown that, as of 2012, only 10% of Saudis undertake some form of technical training.[a] The Kingdom wishes to see this number increase to some 40–45%, a fourfold increase.

[a] City & Guilds: Kingdom of Saudi Arabia invests in skills to boost vocational education provision, 2014.

Since workforce localisation is not the soul remit of IOCs, and IOC's often meet their local content targets through enforcing local content compliance with OFS companies, this section is followed by a more detailed look at the role of OFS companies in developing oil and gas economies in terms of direct employment.

In short, one of the major failings of the current upstream oil and gas model has been the inefficiencies around local recruitment, understanding local skill sets and defining specific areas where local workers can be trained and deployed throughout the life span of a project or asset. Many IOCs have sought to meet their local content commitments through hiring nontechnical, low-skilled staff, such as drivers and security personnel. In addition to this, IOCs have played a juggling game with regards to their local content commitments, often passing the responsibility on to international OFS companies.

The practices mentioned earlier have led to poor workforce recruitment, which has a tendency to hamper the upstream business in many regions, either limiting exploration and drilling activities or causing OFS companies to make severe and drastic changes to their businesses in the wake of smaller contracts and evaporating business.

2.3.3 Workforce: What Would a Model/Approach Driven by Localisation Look Like?

Workforce dynamics need to be balanced, and hiring locals for the sake of meeting contractual arrangements with a host nation has resulted in localisation being relegated to 'corporate social responsibility' and, as a consequence, becoming almost wholly disconnected from core business. As a result, not much has been done to formulate successful localisation strategies – both from a qualitative and quantitative perspective – outside of the work of the previous three books in this series. However, there are some key lessons that can be learnt from looking at the few cases of real, authentic localisation for the sake of improving oil and gas operations from the viewpoint of efficiency, reliability and overall cost reduction. In addition to this, lessons from other intensive industries – where there is an emphasis on engineering and technical competence – can be extracted and employed in the oil and gas industry. In the following sections, we draw on these practices to inform our approach to analysing workforce localisation.

A model or approach driven by localisation would ideally seek to involve OFS companies as early as possible so that direct employment strategies could be developed in collaboration with the IOCs and the host nation government. This would then help to mitigate inefficient hiring of locals and poor deployment of skilled locals in the sector. IOCs could then channel training resources into the right areas to improve local capability, develop legitimate and sustainable succession plans, and create a strong local business that requires minimal technical oversight from head office.

To do this, several things must be addressed, which include the following:

- Corporate attitudes related to the capabilities of local workers in developing countries, which has a significant impact on employment and training of locals, need to change as the perceptions held are generally negative.
- Addressing localisation and the employment of nationals in the sector as early as possible is critical as once operations (and the staffing of those operations) becomes established, it is difficult to adopt new practices.
- Effective use of funds allocated to education and training is vital in assisting IOCs to develop skilled nationals who can be deployed throughout the business. In many cases, huge amounts of money have been spent inefficiently and without proper strategy, resulting in losses for IOCs and poor outcomes for local citizens.
- IOCs and host nation governments often enter into negotiations distrustful of each other. IOC and host nation government relations are too often characterised by mistrust.

Total E&P Angola

Total has significant interests in Africa. Angola presents an interesting case study. Angola's government has sought to use the country's oil sector to improve the lives of locals through both direct and indirect employment. Angola's local content policy includes direct reference to employment requirements, procurement requirements, training requirements, technology transfer requirements, monitoring and enforcement mechanisms and government obligations around the support of the companies' programmes.

Total, in keeping with the company's ongoing commitment to fostering high levels of local content, has ensured its local subsidiary achieves a level of success in terms of employing local people and contracting local suppliers. Since beginning activities in Angola Total has assisted in the development of the local oil industry through technology transfer programmes, direct recruitment and the training of its direct workforce.

A good example of this is Total's work in Block 17's latest project, CLOV. The company made the recruitment and training of locals a key part of the project. Total identified numerous areas where skills and knowledge (alongside technology) could be transferred into its local workforce. This was done through implementing relevant training programmes.

In 2007, Total E&P Angola made an 'Angolanisation' charter, which is testament to, and a guide for, continued workforce localisation in Angola. By the end of 2009, 950 of the 1300 employees (73%) of Total's Angolan subsidiary were nationals, including some 350 managers. Over 50% of those workers were recruited between 2006 and 2009.

Total's commitment to local recruitment for its direct workforce remains part of its recruitment policy. The company maintains strong internal training capacities to maintain high levels of worker competence.

At its core, a model driven by localisation would seek to strategise localisation from the very beginning. In 2016 the OFS industry was concerned that falling profits due to cost-cutting measures by IOCs was causing irreversible harm to the OFS industry. The CEO of Schlumberger commented on this, stating that the 'savings' do not really add up to real reductions in the cost of producing a barrel of oil:

> *The apparent cost reductions seen by the operators over the past 18 months are not linked to a general improvement in efficiency in the service industry. They are simply a result of service-pricing concessions as activity levels have dropped by 40-50 percent, and most service companies are now fighting for survival with both negative earnings and cash flow.*

Schlumberger CEO, Paul Kibsgaard

OFS companies, as a consequence of low oil prices throughout 2015 and 2016, reportedly wanted to be included in discussions earlier. These companies would like to see a new model, whereby projects can be delivered on time, on budget and in line with production targets. This, according to Schlumberger, would involve closer collaboration between OFS companies and oil companies from the start, rather than the OFS companies being brought in at a later point and asked to fulfil a predefined purpose, which often results in higher costs and less efficient operations.

The call for more collaboration and better strategising is perhaps the most crucial point to a model driven by localisation. If localisation targets can be set early on, through effective collaboration between host governments, IOCs and OFS companies, then a strategy focussed around business operations and the life cycle of projects can be developed. This will then have a bigger impact for all parties: IOCs who need to reduce costs, OFS companies who usually inherit IOC localisation targets and are responsible for providing the bulk of the technical workforce, and host nation governments who need to see job creation and economic growth.

Oil and gas companies must also take into consideration how individuals in direct employment are provided with education and training opportunities, in both technical and nontechnical areas of their professional career. Alongside this, oil companies need to measure and provide information related to the progression of nationals in their local subsidiaries or within the core of the organisation's presence in-country. At the earliest stages of establishing their presence in a developing economy, there should be an adaptable succession plan in place, and this will support the localisation of the direct workforce. If a succession plan is in place, IOCs can place key performance indicators or KPIs in foreign workers' targets to ensure nationals receive mentoring and training.

2.3.4 Workforce: How to Approach an Analysis of Workforce Localisation

In this section, we will look at a number of approaches that can help us understand localisation from the viewpoint of direct employment in the oil and gas sector. This section will take into account the following subareas:

1. How to approach an analysis of workforce localisation from the company perspective
2. How to create a gap analysis and a key set of objectives

Approach 1: Analysing the Number of Nationals Employed

This approach is predicated on a head count of nationals employed by the IOC, either through local subsidiaries or directly in the core business. The approach is relatively simple to administer when evaluating a particular project/operation/company's level of localisation and provides a clear and easily understood way of assessing the level of localisation and the rate at which local professionals are being absorbed into the industry. It may also help to identify any skills gaps or employment approaches that work against promoting local participation in the sector directly. Another benefit to using this approach is that it relies on existing information already collected by companies' human resources departments as part of normal business operations and reporting requirements to governments. Reliability-of-data and simplicity-to-administer provide a sense of quality and reliability of the overall assessment and the development of a strategy. The importance of these two benefits cannot be underestimated.

In essence, this approach consists of quantifying the head count of nationals employed by the company in a given region or on a specific project and comparing this to the number of expats employed. This straight comparison allows for an expat-to-local ratio to be generated, from which a very simple, although direct, evaluation of local employment can be made. It is done by measuring the number of nationals employed as a proportion of total full-time equivalent employees.

This metric is often used, particularly by governments, to measure the extent to which localisation is occurring among the directly employed workforce of oil companies and OFS companies. As an example of this, Angola's laws prohibit companies from employing nonnationals unless their workforce is 70% local already. Additionally, this metric can be broken down into the proportion of nationals in professional positions against nonprofessional staff positions according to how the company categorises its job roles. Generally, this is reported in terms of supervisory, skilled, semiskilled and unskilled positions.

In some cases PSAs seek to measure a change in the proportion of locals to expats over time and across different levels of the businesses employment structure. This approach builds in expectations to increase the ratio of locals working directly in the oil and gas sector. For the Shah Deniz Prospective Area in the Azerbaijan Sector of the Caspian, the PSA measures national content in the workforce using the following metrics:

- Prior to commencement of development: professionals 30–50%; nonprofessionals 70%
- Upon commencement of petroleum production: professionals 70%; nonprofessionals 85%
- Five years after commencement of petroleum production: professionals 90%; nonprofessionals 95%

Approach 2: Wages Paid to Nationals

An alternative to measuring the head count of nationals in the workforce or in the execution of a particular contract is to report the salaries or wages paid by the company to national employees and other payroll staff. This is often measured as the gross wage or gross salary paid to and on behalf of nationals as a percentage of total gross salaries paid to all workers. The metric captures base salary and associated social taxes paid by the employee and employer, as well as employee benefits and expenses, including pension contributions, housing, personal and vehicle allowances and bonuses. A similar metric uses taxable salary, that is, gross salary, less employer's contributions and expenses, as this information is usually more readily available from companies' finance departments. Measuring local content by head count or by wages can generate very different results. Below we provide an analysis of how these two important approaches to measuring local employment compare and how each approach can provide different answers.

Consider this example: local content measured as the proportion of nationals in the in-country workforce is at 95%, while local content measured as the proportion of total gross wages paid to nationals is 61%. This would likely be the case if the wages-based metric was applied to measuring the local content contribution of a company that employs expatriate labour in its local workforce in senior management and high-paying technical positions, as often happens especially where these skills are not available in the country of operation. It is not unusual for the head count of national citizens in an oil- and gas-operating company to be around 90% of the total in-country workforce, and for the majority of them to be in the technical, semiskilled and manual categories, resulting in a much lower share of national salaries to total salaries paid by the company. This is especially true in countries with low economic development and less mature petroleum sectors. Both the head count and gross wage metrics are relatively simple to administer, as they are based on information already collected by companies' human resource departments and finance departments as part of normal business operations and reporting needs. Confidence in the reliability of this data should also be reasonably high. Simplicity-to-administer and confidence-in-the-data are important attributes for metrics and should be carefully considered in defining reporting standards or policies.

In addition, regulators and companies should choose the metric that more closely measures the desired outcome. For example, the percentage of national participation in the workforce of a company does not provide information on the extent to which nationals are able to progress in terms of career development, or information on the magnitude of local employment share of benefits. For this purpose, the share of national wages as a proportion of total wages would be a much more informative metric. But this too can be misleading. Although expatriate base salaries are often paid into foreign bank accounts, and thus accumulate overseas, expatriate allowances for housing, transport and personal expenses, as well as expatriate social taxes, will mostly accrue to the domestic economy. These expenditures should not be underestimated.

Below we can see how the number of nationals employed by the oil company compares (*note: for the purposes of illustrating the different results provided by a head count versus wages approach, the examples below use a fictional company with our information drawn from an amalgamation of research data):

Company #A	
Annual labour in man-hours	200,000
Total salaries and wages in US$	3,000,000
Foreign workers annual man-hours	7,000
Local workers in management position: annual man-hours	18,000
Foreign workers in management position: annual man-hours	7,000
Workers in non-management position: annual man-hours	
Foreign workers in non-mangement position: annual man-hours	
Total salaries in US$ paid to all foreign workers in management position	1,550,000
Total wages (excluding expenses) in US$ paid to all foreign workers in non-management position	
--	----------------------------
------------------	----
Nationals in total workforce	95%
Nationals in management positions	50%
Nationals in non-management positions	100%

In our example, we can say company #A has 100 employees, 5 of which are expats working in management positions. That means the total number of local workers, in both management and nonmanagement positions is 95 people. The breakdown of nationals is 5 in management positions and 90 in nonmanagement roles.

5 expat mangers = $310,000 annual salary (total $1,550,000 annually).

Approach 3: Nationals in Contract

The proportion of total man-hours of work undertaken by national citizens in the execution of a particular contract is another way to measure the proportion of nationals in the workforce by head count. This type of metric measures the number of nationals actually involved in a particular job. Thus it avoids the situation whereby a high level of local content is assigned to a segment of procurement expenditure because of a high national head count in the service company, when actually this figure depends heavily on nationals working in administrative positions, that is, not directly involved in the contracted work.

How to Create a Gap Analysis and a Key Set of Objectives

Once the methodology for analysing and assessing localisation of the direct workforce is established, and once an initial assessment has been undertaken, an IOC, NOC, company or government is in a place where an accurate gap analysis can be completed.

A gap analysis can enable all parties involved in the localisation process to achieve several aims:

- Understand the skills required within a certain time frame (between the point of establishing an asset and running that asset for a set number of years);
- Understand what skills are missing in the local workforce and why these gaps exist;
- Identify immediate job profiles that can be wholly or partially filled by locals;
- Produce a pipeline map of local talent that can be utilised in the short to medium term;
- Create a road map of investment to develop the pipeline of talent according to the business needs within the context of a carefully managed approach utilising relationships between industry, the public sector and the education/ training sector;
- Develop a succession plan that will work to set specific goals around the progression of locals within the industry in a sustainable manner that also takes into consideration the needs of the operators in-country;
- Utilise in-house training methods to enhance local workers, bringing to bear international experience in interventional training methods and *rapid training*.

Using the information in this chapter we can begin to construct a gap analysis relating to the direct workforce employed by both IOCs and OFS companies. The gap analysis should start with understanding the current level of localisation and measuring this against some key metrics:

- National citizens employed, represented as a percentage of the total IOC workforce on the basis of a head count;
- National citizens in senior, skilled and supervisory positions on the basis of a head count according to job type;
- Number of man-hours worked by nationals on an annual basis;
- Value of wages paid to nationals as a percentage of total wages paid to workers on the given project on an annual basis;
- Breakdown of wages on an annual basis according to job type, comparing international workers' wages to that of nationals'.

Stage 1 of this process will focus on workforce analysis: this analysis, according to the above metrics, should be conducted for the direct workforce employed by the IOC(s) as well as any international or foreign-owned OFS companies that are engaged by oil and gas actors to fulfil tasks on the ground. This would be broken down accordingly:

Stage 2 of the process will involve a localisation viability assessment: after collecting this data an assessment of the potential for further localisation can be made. This includes understanding the potential savings to be made from an

	IOC	OFS
National citizens employed, represnted as a percentage of the total IOC workforce on the basis of a head count.		
National citizens in senior, skilled and supervisory positions on the basis of a head count according to job type		
Number of man hours worked by nationals on an annual basis		
Value of wages paid to nationals and a percentage of total wages paid to workers on the given project on an annual basis		
Break down of wages on an annual basis according to job type, comparing international workers' wages to nationals'		

increasingly localised workforce and any potential benefits host nation governments offer to IOCs who manage to facilitate localisation through skills transfer.

Stage 3 will require an analysis of host nation government requirements related to localisation. If a certain amount of money must be spent to facilitate localisation and the succession of locals into oil and gas sector jobs over a given period of time, this should be integrated into the gap analysis. This analysis should include an analysis of national development priorities.

Stage 4 can take place once Stages 1–3 are completed and involves designing a strategy that will work towards meeting the gaps identified through completion of the first three stages. The strategy should seek to do the following:

- Address the need to lower costs of operations related to the hiring or contracting of trained, field-ready professionals.
- Address any concerns related to the competencies of national staff. This includes health, safety and environment (HSE) requirements and core technical competencies.
- Address the need to meet any government requirements related to localisation. This could include succession planning and filling higher tier job profiles in middle and senior management roles.
- Include nontechnical training strategies to improve localisation so that technical staff can gradually move into management roles within the local business.
- Address any foreseen issues related to enculturating the company's current methods and systems for training technical staff and assessing competencies (e.g., complex language in competency assurance systems that leads to confusion among national staff).
- Create a pipeline of talent to meet the future needs of the industry. This means mapping the capacity of local education institutions so that investments in local training and education providers can be harnessed to full effect, allowing

the IOCs and OFS companies to maximise their investments (which are usually stipulated in contractual agreements with the government or in local content policies).

- Address any issues that may hinder a company's ability to recruit locally, such as recruiting practices that are alien to nationals.
- Ideally, the strategy would look at localisation in 3-year increments, providing time for the company to reflect and adjust the strategy accordingly.

Additional information to support the gap analysis and the development of a localisation strategy include, but may not be limited to, the following:

- A local capabilities study, which would help to inform the gap analysis at Stage 3. This looks at existing and potential capabilities, which is particularly important when operating in a new location. The study can draw on local expertise and the use of market intelligence. It is advised that international partners as well as local partners are engaged together to increase the validity of this local capabilities study. This will not only help to map the local skills, but provide an international context.
- Conducting an 'Environmental and Social Risk Assessment' would be hugely beneficial to this process. An assessment of this nature should seek to highlight any issues related to the ability of local contractors to deliver to international HSE standards. An assessment of this kind can also serve to demystify certain cultural practices that may impact on the company's ability to engage nationals for work in the oil and gas industry. Assessing the social factors is hugely important and, if neglected, can severely impact on the localisation strategy.
- Cost–benefit analysis is a critical element to any localisation strategy. As mentioned previously, the changing model of the upstream industry must increasingly focus on the training and development of nationals if IOCs are to create savings and cut costs related to the production of oil and gas. The localisation strategy, which should be the outcome of the final stage in the above process, should focus heavily on how local recruitment, informed by a gap analysis, a local capabilities study and an environmental and social risk assessment, can reduce costs. Investing in the talent pipeline (e.g., investments in – *not donations to* – local education institutions) can be made strategically to reduce costs related to staffing and recruitment. The cost–benefit analysis should quantify and analyse the different levels of benefits, costs and risks tied into local employment. At this stage, it is important to take a long-term view of cost–benefits and not look solely at the immediate ways in which costs can be reduced.
- Another informative study would be a 'Barriers Analysis'. This would help to identify any barriers the localisation strategy may encounter. As stated above, it is best for the localisation strategy to look at localisation in 3-year increments. An analysis of the barriers that could potentially limit or impact on localisation efforts would ideally looks at these barriers within the same time frame, addressing specific problems the company could encounter. An assessment of these barriers helps to identify them earlier and thereby create mechanisms to counter them.

- Some form of infrastructure analysis can contribute greatly. Localisation depends on existing infrastructure. Local suppliers might be limited by the lack of local infrastructure, which may in turn render them incapable of participating in contracts that would engage nationals directly in the oil and gas industry. A lack of infrastructure can be a barrier to the provision of local services. A lack of important infrastructure can also increase the costs of doing business for the IOC or OFS company. Unreliable power supplies may force companies and suppliers to purchase generators and fuel. A lack of shipyards and ports may limit local companies from participating to offshore operations and other areas integral to the core operations of operators. A lack of transport links can make the delivery of core components a problem.

In the model presented here, we believe that employing several studies, those described and listed earlier, increases the reliability of the strategy and enables successful localisation with real cost–benefit outcomes for the IOCs and operators engaged in local operations. The early stages of moving into a new area of operation must be characterised by a deep and exhaustive understanding of local workforce dynamics, skills, education and the availability of local workers for immediate assimilation into oil and gas employment, both at the core technical level and in the business and administrative areas. Additionally, this sort of analysis does not only apply to IOCs but the international OFS companies who currently provide the majority of workers at an asset and provide IOCs with the engineering and technical support required to successfully exploit an oil or gas field.

Due to the importance of understanding the OFS industry's role in localisation, the next section will deal directly with the OFS industry, looking particularly at how the OFS industry has become so important to IOCs and why early involvement of the OFS industry will enable higher levels of localisation that will support IOCs in meeting cost targets and ensure the industry is achieving the national development targets of the host nation government.

2.3.5 Workforce: Understanding the Need for Localisation in OFS Industry

The global OFS sector has enjoyed a sustained period of quiet progress. While large OFS companies rarely get the media attention of the IOCs – which can work in their favour when catastrophic accidents occur – OFS companies have been the driving power behind the recent expansion of exploration, drilling and production activities across the globe and their specialised services are sought by IOCs and NOCs alike. As oil prices soared, and oil and gas companies could afford to spend more, OFS companies directly benefited from the abundant high-value contracts on offer. This allowed OFS companies to expand their businesses and move to the forefront of technological and knowledge development. OFS companies have invested, and continue to invest, huge amounts of capital in the development of their staff and their technology. Since OFS companies do not hold any oil and gas production assets, their assets are their staff. This has made the OFS industry indispensable in the modern hydrocarbon sector. These companies offer everything from training services to horizontal drilling, EOR to 4D seismic services. Between 2003 and 2013, global oil reserves increased 27% to 1.7 trillion barrels (according to a report by PwC). This increase was significantly attributable to the role OFS companies played in technology and skills development, something that has greatly enhanced the ability of IOCs and NOCs to tap into previously inaccessible reserves. Additionally, for the same reasons, gas reserves grew 19% to 186 trillion cubic metres during the same period. All of this contributed to the tremendous growth of the OFS sector. Consequently, and perhaps unsurprisingly, OFS companies have, in recent years, outperformed IOCs and NOCs in terms of revenue, market capitalisation and stock price.

One of the most important aspects of this rapid growth of the OFS industry is the fact that now these companies typically invest more in research and development than most IOCs tend to do. According to reports issued by IOCs and data analysed by Strategy&, the largest OFS companies set aside an average of 0.7% of sales for research and development, compared to some 0.4% of super majors (both IOCs and major NOCs).

In order For OFS companies to survive market and business disruption they should do the following (information sourced from PwC):

Contracting and pricing:
- realign pricing and risk sharing to echo current conditions;
- seek more collaborative development/performance contracts and
- offer end to end and turnkey services contracts.

Cost management:
- negotiate input prices downwards to reflect reduced demand and rising costs;
- restructure internal costs with flexible operating models and cost rationalisation and
- localise operations wherever applicable to utilise local labour and supply chains and reduce costs.

Integrated offerings:
- horizontally integrate across processes/segments;
- offer integrated systems versus components to capture more value and
- provide process and systems development to enhance customer value.

Portfolio optimisation:
- shed marginal or incoherent businesses or products;
- acquire businesses and products to fill in integrated offerings and
- exit markets and areas that are subscale or poor performers.

According to analysis conducted by PwC, it can be clearly observed that OFS firms will need to secure themselves against a low oil price and generally low spending from oil and gas companies who are becoming more and more risk averse and are veering away from high levels of spending with international contractors. OFS companies will, very quickly, need to adapt to these changing circumstances by offering a greater number of services and be more willing to share the costs and – most importantly – the profits. We have seen the use of performance-based contracts becoming increasingly popular among OFS companies and their clients. Contracts of this sort allow for a greater degree of risk sharing, particularly the financial risk. Additionally, performance-based contracts offer the real possibility for OFS companies to improve their profits on the back of a highly successful E&P operations. The trade here is the lower upfront costs to the client but bigger bonus payouts in the event of a successful venture.

The modern operating environment is, undoubtably, putting a strain on OFS companies. However, these companies simply cannot afford to not renegotiate their costs and they cannot depend on the now very scarce oil boom contracts. Now more is expected for less from OFS companies. The OFS industry will need to involve itself earlier in discussions, no longer playing the role of a contractor but seeking an intimate, cost-effective partnership with IOCs and NOCs in whatever region they are operating.

In the case of NOCs, the OFS industry will need to forge the same partnerships, based on cost-sharing. This will involve, in part, a change in the approach NOCs take, where they use OFS companies to supply them with the required technology for production. Forging results-linked partnerships will help OFS companies and, in the long run, create value for the host nation.

OFS companies, who ultimately receive responsibility for meeting their clients' (the IOCs) local content targets, can stand to profit immensely through strategic localisation and knowledge transfer. As OFS companies take on greater responsibility for *managing* the asset and not just running it, the need to reduce costs, increase efficiency and see greater profits come out of the asset will be integral to the OFS industry as it transitions into a new risk-sharing model. It will only survive this transition if it learns, as it has been doing, the importance of localisation and places this at the heart of the business model. Everything we have discussed in relation to the need for IOCs to embrace localisation applies in full to OFS companies.

This presents an opportunity to work with players in emerging economies to share the risk and reduce costs through training and employing locals. However, the real impact comes in the form of greater partnerships with IOCs and NOCs, whereby OFS companies can work with local engineering companies, sharing contracts with these companies, improving their scalability and facilitating technology and knowledge transfer.

IOCs and NOCs, appreciating the complexity of modern oil and gas extraction and production and acknowledging their reliance on the skills, expertise and technologies of OFS companies, can and should seek to engage OFS companies in the early stages of extraction. As mentioned previously, the opportunities for cost-reduction through increased localisation offer significant opportunities for all parties.

2.3.6 Workforce: Concluding Remarks

In this section, we have looked at the current model for the upstream oil and gas sector and discussed some of the most important factors and pressures that are forcing IOCs, NOCs and OFS companies to adjust their traditional approach to securing a workforce for exploration and production activities. We have also looked at how an evolved approach to localisation, one that sees the cost–benefits instead of only the social responsibility component, can improve performance and, overall, reduce costs related to staffing and recruitment. Finally we have looked at localisation, how to approach an analysis of localisation, and how to structure a business strategy around localisation for the modern upstream model.

It is worth stating that, when looking to localise your direct workforce, and analysing the cost–benefits of doing so, it helps to keep in mind a simple question: where is the tipping point at which localisation becomes a cost-burden and stops being a cost–benefit? For this reason, any strategy must be adaptable and flexible enough to meet identified and unforeseen barriers that could manifest at any point.

In this analysis we have seen how local companies can also be engaged to participate in direct technical work at an asset. This provides localisation from two view points: first there is the local workers aspect, and second there is the supply chain component, which is arguably more important for long-term, meaningful localisation than direct employment. However, for the sake of this section we have limited our scope to assessing several key localisation indicators and how these are impacted by the number and salary of local workers, regardless of whether these workers are found in the IOC, the NOC, the OFS industry or in a local company supplying workers for core technical operations. This reflects the current model of the oil and gas upstream industry, where direct employment in the sector cannot be simply defined in terms of how many nationals the IOC has employed. Rather, we need a wider frame of reference for direct employment and a more serious analysis that views local employment from a variety of angles, including job type, salary, and man-hours.

2.4 THE SUPPLY CHAIN COMPONENT

As the global oil and gas industry moves towards a future of lower prices and tighter margins, the need to shorten supply chains and reduce costs related to the procurement of goods and services is becoming increasingly important.

Localising the supply chain can give IOCs significant cost reductions, and having strong local supply chains with locally owned vendors can create other benefits around ease of doing business. Furthermore, the impact of supply chain opportunities on employment and on the wider economy can be profound for oil and gas nations, particularly once the operations are up and running and the direct employment opportunities with oil companies and OFS companies is limited. However, IOCs do face serious challenges when it comes to localising their supply chains and procuring goods and services from local vendors.

In many emerging nations, manufacturing and fabrication capacities are limited. In terms of services, there are generally lower levels of services in terms of scalability, primarily due to local vendors being informally employed tradesmen within a larger informal technical workforce. Consequently welders, pipe fitters, bricklayers, etc., are – as a general rule – informal workers lacking formal qualifications and working from job to job. This can create significant issues related to finding suppliers who can deliver services at the scale required by IOCs.

These difficulties however significant are not entirely insurmountable. Provided that a 'Localisation of Procurement' strategy is developed early on, and this strategy remains adaptable and produces the 'local content' sought by governments as well as accomplishing the business objectives of the IOCs.

In this section, we will discuss the current model of procurement in the international oil and gas sector, which is dominated by international vendors who either operate under their international brand or register in-country for legal purposes. Additionally, we will examine what a modern model of supply chain localisation would look like. We will then present a way to measure supply chain localisation and – just as importantly – the how to analyse investments in local supplier development.

2.4.1 Supply Chain: The Current Model

The oil and gas industry faces constant challenges related to the supply of important goods and services, and balancing this supply with demand driven by a constantly shifting commodity market. When prices are high, oil companies have relied on large international service provides to meet their core technical needs. This has made the OFS industry a highly lucrative area (something we addressed in the previous section). However, when oil prices fall, oil and gas companies begin to scrutinise their existing contracts with OFS companies, looking for areas where services can be cut back, or cut out.

There are three main 'types' that have been identified in terms of how the current IOC procurement model works. Looking at these types serves to demonstrate the relationship between IOCs and their supply chains (contractors). The models below do not take into account the subcontracting of suppliers undertaken by those contracted to IOCs.

IOC X operates in Iraq and, across its assets, its relationship with contractors has the following features:

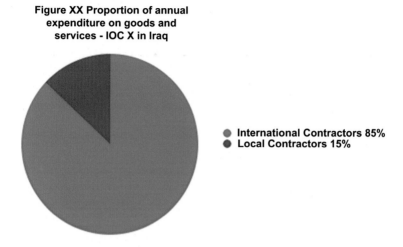

Figure XX Proportion of annual expenditure on goods and services - IOC X in Iraq

● **International Contractors 85%**
● **Local Contractors 15%**

As we can see from the above chart, the current model for procurement of goods and services among IOCs (in many parts of the oil and gas-producing world) is characterised by overreliance on international suppliers. There is typically very little local companies (that is, companies registered locally with 'local' owners) can do to increase their role in the supplying of goods and services to the industry. Another key characteristic of the relationship between local companies and an internationally dominated oil sector is what sort of goods and services are generally supplied to the industry. Most often, IOCs and international OFS companies will engage local companies to provide goods and services that are not core to the technical activities of industry, but sit more suitably in the business and miscellaneous needs of the industry. An example of this would be Nigeria, where very few companies contribute to the supply of goods and services related to core technical areas, engineering, fabrication and manufacturing. Instead, local companies are sought to fulfil the catering, security service needs and to provide basic supplies.

In Section 2.2.1, we put the emphasis on understanding the workforce integral to the running of the oil and gas sector: this means all people directly employed either by IOCs or by OFS companies across the many areas and job profiles that exist in the sector. In this section, we will look at the oil and gas supply chain and provide a categorisation of the different segments of the supply chain. One limitation to our study in this area is the degree to which companies decide to internalise certain services, especially professional services such as accounting and legal services. A degree of flexibility must be applied when examining the supply chain of the industry, and this involves analysing the international operator on a company by company basis and understanding the individual operator procurement needs. That said, here is a broad list of categories – based on our research and discussions with industry leaders – of supply chain areas[1]:

- Legal services
- Accounting services
- Research and development
- Engineering and CAD services
- Banking
- Insurance
- Hotel and hospitality services
- Catering services
- Freight services (by water)
- Freight services (by road)
- Passenger transport services
- Charter services
- Warehousing and storage
- Cargo handling
- Motor vehicle maintenance
- Employment agencies and recruitment/HR consulting services
- Travel agencies
- Security services
- Facilities management
- Cleaning services
- Office administration
- Medical
- Data processing
- Web development
- Telecommunication and networking services
- Computer programming
- Site prep services
- Foundations/masonry/scaffolding
- Crane operation services
- Construction – roads, ports, buildings and essential infrastructure
- Docks-related services
- Plumbing
- Electrician
- Concrete forming services
- Water supply services
- Waste management services (nonhazardous)
- Training and development
- Power supply operations

---

1. This should not, at this stage, be considered an exact and unchangeable list of supply chain areas for the oil and gas industry.

- Heavy equipment maintenance
- Iron and steel manufacturing
- Steel/pipe casting/shaping/forming
- Fabricated metals products manufacturing
- Metal frameworks manufacturing
- Industrial machinery repair
- Pumps/compressors/valves manufacturing
- Industrial gases supplies
- Sanitation and sewerage services
- Hazardous waste treatment and disposal
- Environmental monitoring
- Pollution control

In the current model for procurement of goods and services, IOCs, NOCs and OFS companies often fail to fully understand the capacity of local companies to meet their procurement needs. It is also commonplace that procurement processes and contractual issues mitigate against local companies being able to bid for contracts. Whilst we recognise that the use of local companies in fulfilling supply chain needs varies from country to country and company to company, the current model typically fails to fully engage with local suppliers, and as a consequence the wider economic benefits of oil and gas production are not felt locally.

2.4.2 Supply Chain: Failures of This Approach

This approach, where locally owned companies only take a minimal role in terms of meeting the procurement requirements of IOCs, fails for several reasons. While there are numerous corporate social responsibility issues with this approach – and while the approach fails to adequately spread the benefits of oil and gas production amongst the local supply chain – perhaps the most pressing reason this approach needs to change is that contracting a majority of services out to international providers is neither efficient nor cost-effective. Even though international supply chain companies are willing to drop their prices and renegotiate contracts, there are many elements of the required supply chain where IOCs and international operators could opt for in-country support. These include, but are not limited to technical services, fabrication, engineering, transport and logistics, research and development and construction.

Additionally, supporting the development of local suppliers – and of the functioning of the supply chain as a whole – so that procurement can be more concentrated at a local or regional level, does more than just help to reduce costs in the long term. It provides IOCs with a 'moral license' to operate. In numerous case studies where IOCs have failed to engage, or even simply ignored, locally owned suppliers, local people have turned against the IOCs, demonstrating their dissatisfaction and creating real and sustained problems for operators. There are two clear (and opposing) examples, in Nigeria and Libya, where this has been amply demonstrated (see the below boxes).

Nigeria

Nigeria is often cited as an example of a country that has suffered the 'resource curse' where commercial level of oil and gas production do not translate into national development, but rather becomes the driving factor behind lacklustre economic growth with severe social and environmental consequences.

In Nigeria, the oil and gas sector is dominated by IOCs who have, for many years, protected their own interests at the expense of local communities who have not only seen very few opportunities arise from a growing E&P sector, but have experienced the rise of crime, corruption, social unrest and that can be directly traced back to the mismanagement of the oil sector.

The local supply chain has seen very little stimulation, despite a fairly comprehensive local content policy, which stipulates a high proportion of Nigerian companies are contracted to provide products and services. Local ownership of operations has been a major component of local content policy in Nigeria, which would, it was thought, help local supply chain companies benefit from the opportunities offered up by international companies. This emphasis on locally owned businesses and their involvement in the oil sector was present throughout the oil industry, with contracts and subcontracts having specific stipulations for how local companies should be engaged as part of any subcontracts issued by the IOCs.

While obviously ambitious – and arguably impossible for a globalised industry – these local content rules had little effect on how international businesses conducted themselves in Nigeria. This was due to a number of factors and was by no means the result of international operators alone.

A practice known as 'fronting' became rife. Fronting allowed international companies to sidestep local content by becoming subcontractors to local companies that had little to no expertise to deliver technical and engineering services, or other services required by IOCs. This effectively meant that local Nigerian-owned companies were only a front for international companies and therefore very little value was added to the country.

Nigeria's example serves to demonstrate how financial mismanagement and a lack of interest in the development of local suppliers cannot only hinder a country's development, but can create a situation whereby oil and gas production leads to declining conditions for locals, their communities and the country as a whole. Locals have protested against IOCs, causing companies to suspend their operations due to safety concerns.

Nigeria is moving forward, however. In the past, Nigeria was buying goods and services for resale to the oil sector, whereas now the focus is moving towards added value and value retention through in-country manufacturing and fabrication of goods. It is estimated that capital flight has been in excess of US$380 billion. By 2020, Nigeria hopes to be a regional hub for services in the oil and gas supply chain, retain some US$191 billion and create 300,000 jobs through supply chain development.[a]

---

[a] http://www.ilo.org/wcmsp5/groups/public/---africa/---ro-addis_ababa/---ilo-abuja/documents/publication/wcms_444850.pdf.
Information from Getenergy Guides Volume 3.

Libya

At the beginning of the 21st century, Wintershall operated its assets and worked with the government of Libya to secure job growth for nationals and invest in local suppliers. Wintershall worked to create a sustainable model of job creation, both directly and indirectly, through recruitment and the contracting of local and nationally owned companies to meet supply chain requirements. While the political and security landscape in Libya was riddled with uncertainty, mismanagement and very much subject to the whims of an authoritarian dictator whose policies were liable to change at any given moment, Wintershall earned the respect of local people and communities who saw the company as a stable force in the oil industry and one that was committed to the betterment of Libyan society.

In 2011, in the midst of a revolution that would eventually see the overthrow of then Libyan President Muammar Gaddafi, Wintershall was faced with ceasing oil production or endangering its staff. However, the company reported that during this time of armed conflict and sectarian strife it experienced only a marginal decline in its production figures. This was largely due to the fact that Libyans respected the company and sought to protect it and even isolate it from the conflict as it was in their own interests to do this. This demonstrates how investing in local people - and engaging with the local supply chain - can build loyalty and, ultimately, instill resilience into the business.

Information from Getenergy Guides Volume 2.

What we learn from these case studies is that failure to engage effectively with the local supply chain not only hinders local economic development but also generates real challenges for international companies in a given country. By not utilising local suppliers effectively, oil and gas operators create tensions in local communities, damage their reputation with regional and national governments, and critically fail to achieve the economic benefits of sourcing goods and services locally.

2.4.3 Supply Chain: An Approach Driven by Localisation

As discussed in the previous section, any new model that addresses the localisation of the supply chain must take into consideration the changing dynamics of oil and gas exploration and production. There are a multitude of business and technical realities that must be addressed. By way of a summary of these changing dynamics, here are some of the most pressing in today's economy:

- There are now fewer specialists and engineers serving the industry
- Local content policies and government regulation of the industry are both evolving rapidly
- Lower oil prices are the defining context for the industry as a whole

- There is now process of downsizing, with large international companies employing fewer people

Arguably, the most important consideration for oil and gas companies is how they procure goods and services that are integral to their core business operations. Taking the most critical components of an IOC's operations, that is those services critical to the running of the business from a technical and commercial perspective, oil and gas companies have been extraordinarily reluctant to work with their supply chain to enable efficient collaboration that could lead to significant savings for the oil company while having the added benefit of building local content. Our model of localisation must account for the needs of integral supply chain companies while also meeting the requirements of IOCs in a way that is sustainable (cost-cutting and squeezing OFS and engineering service companies are not sustainable in a high-reliability industry).

Moving forward, the new model for the oil and gas industry should seek to address issues surrounding local procurement and how this can be enabled. While a certain amount of 'local supplier development' is necessary, it is important to acknowledge how investment into the local supply chain can greatly reduce the costs incurred by IOCs who are dependent on specific goods and services. This will mean that IOCs will need to do two things: continue to seek noncritical services from local suppliers and work with international OFS and engineering service companies to support the development of local providers able to supply some of the technical needs of the industry. This will require several things: an intimate understanding the local supply chain and it's capacities to meet the demands of the oil and gas sector, and it will require oil and gas companies to rethink their procurement practices.

2.4.4 Understanding How We Measure Localisation in Terms of the Goods and Services That Are Needed to Support Oil and Gas Operations in Any Given Country

Supporting the business needs of international operators in new oil- and gas-producing countries involves understanding how and where local procurement can take place, where the opportunities exist for locally owned companies and how to measure the extent to which value is being added to the sector. The first step in developing a strategy for procurement localisation is to conduct an analysis of local content in the supply chain, similar to the analysis conducted for direct workforce localisation.

To properly measure local content in terms of supplier procurement, there are several metrics that will first allow us to build the information we need to begin a gap analysis. These are covered below.

Metric	key feature	Information source
Value of contracts awarded to locally-owned/nationally-owned suppliers as % of total spending on suppliers (internationally competitive basis)	Value of contract awarded to local companies	Reports related to contracts awarded
% of tender lists comprising nationally registered suppliers	Successful tender lists	Company reports and company vendor register
No. of contracts awarded to nationally registered vendors on an internationally competitive basis	The number of contracts awarded to local suppliers	Company reports and company vendor register

These are the key metrics for reporting local content and help us to establish an understanding of where local companies are able to contribute and how they can be supported to contribute further.

Results can vary considerably and some more work must be done to define how an accurate local content measurement can be taken. There are two defined approaches to measuring local content, as set out by the World Bank:

- With reference to a classification of local suppliers
- With reference to a classification of local goods and services

While on the surface these approaches seem very similar, they differ greatly when applied to the metrics above. In the first instance, local content will be measured by whether the local supplier is deemed to be 'local', while in the second instance the service or goods must be considered 'local'. For goods to be considered local, it is crucial to determine whether the relevant goods have had value added in-country.

Classifying Local Suppliers

Definitions used by local content regulations and oil and gas companies to classify what is spent on a local supplier usually include one or a combination of the following criteria:

- The address given in vendor registration information
- The address on a purchase order or invoice
- The geographic location of the service being provided or the production of the good being supplied
- The share of equity owned by national citizens (for example, greater than 25, 50 or 100%)
- Whether the supplier is locally incorporated
- Whether the supplier is tax registered in the country of operation, including for withholding tax purposes
- Whether the supplier employs more than a specified percentage of nationals

With regard to regional or community suppliers, more refined definitions can include the following:

- The regional/local address in vendor registration
- The regional/local address on invoice
- Suppliers and contractors who source the majority of materials or labour from the province, district, or local communities located closest to the operation

Angola defines an Angolan company as any company that is legally established or constituted in Angola, has its headquarters in Angola, and at least 51% shares and participation in the company's equity in the company owned by Angolan citizens. While, as another example, Kazakhstan defines a Kazakh company as one that is producing goods in-country or any company that has more than 95% national ownership (Kazakh citizens).

Measuring Expenditure on Local Goods and Services

It is important, as well as understanding whether or not a company is 'local', to understand whether the services or goods are themselves 'local'. From this perspective it is important to use what the World Bank has called 'the rules of origin', which classify a service or good according to the country of its origin. According to the world Bank:

> *These rules are now being applied by some regulators to formulate local content targets and reporting requirements. The implication of this method is that a company will be able to report local content only if value was added in country in order to provide a service or produce a product.*

Measuring Supply Chain Local Content Through Number of Contracts Awarded

In the graph below, we can clearly see how a typical oil and gas company might divide its procurement between international suppliers and local suppliers in a frontier country:

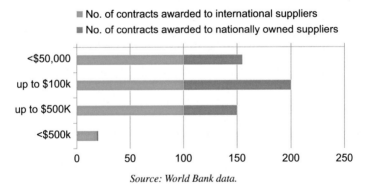

Source: World Bank data.

This demonstrates the extent to which a typical approach to supply chain fulfilment excludes local companies more and more the larger the contract becomes. Although the local capacity of the supply chain is clearly a key factor here, we can also see how much of the revenue that goes into an oil or gas project is likely to end up pouring out of any given country under the current model.

2.4.5 Model for Implementing Supply Chain Localisation

In the following sections, we will lay out an approach to formulating a supply chain localisation strategy. In this strategy, we will look at four overarching stages required to develop the strategy as well as providing some additional studies that can contribute to each stage of the development process.

Stage 1: In the first step it is necessary to understand the supply chain needs and map these needs. This process is a fairly standard procedure and is a business requirement, usually conducted in the early stages of a new project. This involves quantifying the expected demands of the business in relation to a specific project or asset.

Stage 2: In the second step it is necessary to quantify the supply. To do this a supply chain mapping exercise is advised. Market research should be conducted to understand the following about the local supply chain and its vendors:

- Availability of vendors
- The capacity of local vendors to deliver to the scales required by the business
- The quality of output produced by local vendors

Stage 3: Once steps 1 and 2 are complete, a gap analysis should be conducted so that the quantified demands are compared to the quantified supply alongside a market research study that evaluates local vendors.

Stage 4: Once a gap analysis is complete, the final stage is the development of a strategy through which two things can occur with maximum cost–benefits for the business:

1. localised procurement
2. supply chain development

Localised procurement relates to strategies around adjusting the procurement process, shifting away from one vendor to, perhaps, multiple vendors (both local and international) working together to provide goods and services.

Supply chain development relates to the training, knowledge transfer and technology transfer that can be realistically achieved to improve the scalability of local vendors and improve their capacity to meet the business' demand and also achieve the required quality of output.

2.4.6 Supply Chain: What Would be the Benefits to Adopting a Strategic Approach to Localisation in Regard of the Oil and Gas Supply Chain?

Adopting a localisation approach that encompasses a strategy around the localisation of procurement of goods and services has many significant benefits to all concerned:

- Through early identification of local suppliers, vendors can be encouraged to participate actively in the oil and gas industry and this will, in a very short time, create value retention for the country.
- As value retention is nurtured – even in the early stages when local vendor participation may be limited and require changes to procurement procedures – local companies will experience the growth prospects that value retention brings.
- If the proper drivers are in place, and with the help of international operators and effective government policy, value retention can translate into scalable growth for participating local companies.
- Scalability will also require investment in education and training so that new skills can be integrated into the local workforce for the supply chain. In some instances, this can be achieved through partnerships with international companies. This approach involves international companies collaborating with local companies to facilitate knowledge and technology transfer.
- International companies can assist local companies in finding financing options to fund education and training for capacity development.
- Through strengthening the local supply chain, IOCs can significantly reduce their costs and diversify their vendor lists. The introduction of competition locally can help to drive down the costs of goods and services for the IOCs while promoting value retention and value-added production in country.
- By encouraging growth and supply chain capability, oil and gas companies will naturally assist the national development goals the host nation has pinned to its burgeoning hydrocarbon sector. This has a significant impact on the reputation of the IOC and can help to secure their assets in country.

Supporting Supply Chain Development in Emerging Host Nations

In this section, we have so far discussed metrics for looking at how involved local vendors are within the oil and gas industry, and this looks at the number of contracts awarded to local vendors, the value of those contracts when compared to their international counterparts, and other key features relating to understanding the role of local vendors in supporting and supplying the national hydrocarbon sector. However, this does little to evaluate gains made in the productivity of the supply chain's workforce and little to help us to develop those workers and the companies they work for.

Other metrics (sourced from the World Bank and developed by the World Business Council for Sustainable Development) can be used to better understand the progress and development of supply chain vendors. However, these metrics tend to be harder to administer and the data is not always clear. That said, the metrics do provide guidance for understanding how a vendor can be developed and provide some level of insight as to the progress being made. These metrics are included below:

Metric	Key feature	Information source
No. of contracts awarded consequential of supplier support from parent company		Market survey
Value of contracts awarded consequential of supplier support from parent company		Market survey
Amount spent on facilitating skills development		Market survey
Increases in local supplier scalability and capacity e.g. storage capacity, crane lift maximum weights, etc.		Market survey
Improvements in local vendors' HSE performance	HSE	Market survey
Improvements in accident reports, first aid cases, etc.	HSE	Market survey
Individual vendor growth in terms of skilled workers		
Levels of capital investment committed to capacity building of local vendors		
Supply chain resilience: coping with disruptions		

These metrics have been selected from the World Business Council for Sustainable Development, but a more comprehensive list does exist (see WBCSD, 2012) and readers are encouraged to use the metrics above and the metrics contained in WBCSD documents to formulate their own approach to understanding supply chain development and to track the progress of vendors they are working with. Flexibility must be maintained since, depending on the circumstances and the area of operation, having a fixed, rigid formula may highlight areas of development that have been targeted in discussions between IOCs and their local suppliers. In addition to these metrics, IOCs and international operators may want to use metrics related to the competitiveness of the local vendors' workforce.

Oil and gas companies should expect that local vendors who are involved in supplier development programmes provide savings to the company concerned that are at least equal to the initial capital investment. This is an important point and one that should not be ignored. Most companies, according to the World Bank, do not employ the use of metrics that measure the competitiveness of local vendors, nor do they use metrics that measure the sustainability of their businesses. Instead, most international operators prefer to measure local content itself and focus on the level of in-country value retention. According to the studies conducted in this area, measuring local content in terms of supplier development may add an additional 15% of local content (value retention). The World Bank provides a good example of this relating to BP Migas, formerly the oil and gas regulator for Indonesia:

> *[In] Indonesia, where BP Migas—the oil and gas sector regulator—has introduced regulations that reward local suppliers who show expenditure on developing the environmental management and health and safety standards of local material and equipment suppliers, or expenditure on promoting partnership programs to build the capability of local enterprises lower down the supply chain. Documenting local content development in this manner may contribute up to an additional percent of local content, which is then applied to the bid price to afford an advantage in the process of awarding contracts.[2]*

The above metrics relate to 'input', that is to say the level of assistance and financial contributions a local supplier receives. On the other side of this are the output metrics, which are rarely employed, and involve tracking the development of the vendor(s) over time. Typically this will look at improvements in delivery time, improvements in the quality/suitability of goods and services, volume capacity, purchase of technologies and HSE performance. Of particular importance from the output perspective are the metrics related to contracts won by the local vendor with the IOC or international operator and how the revenue of local vendors has grown over time. These key indicators are extremely important in determining real supply chain development and are, arguably, the most important metrics to keep track of in any localisation strategy relating to the supply chain.

2.4.7 Supply Chain: The Challenges/Barriers to Adopting a Localisation Approach for Local Procurement

In adopting a more proactive and strategic model of localisation for the procurement of goods and services, we recognise that a number of key challenges exist. These include the following:

- The first – and arguably the most difficult challenge to overcome – is the notion that local supply chains in developing economies are incapable of meeting the needs of multinational companies. This thinking is largely a result of corporate management culture. The challenge – some have argued – is only in the heads of those who believe it and buy into the false narrative of 'local incapability'.
- The idea that local craftsmen and artisan workers are unable to replicate the standards of work of their counterparts in developed economies is generally a fallacy. In instances where craftsmen from the UK were sent to emerging economies to train artisans, the craftsmen unanimously agreed that local

---

2. World Bank.

artisans were capable and skilled, achieving a high standard of results even though they were often required to use older, less efficient tools. Despite this, these prejudices endure.

- Measuring value retention is inherently difficult and therefore undermines any efforts to quantify supply chain localisation.
- There are no guarantees that local suppliers will deliver on time and to specification, and it can be challenging to know the quality or capacity of a local supplier before you have used them.
- Misunderstanding the capabilities of local vendors can create additional costs if those suppliers are unable or unwilling to meet deadlines or work to a certain level of quality.
- Quality assurance is always a real and important issue when sourcing from local suppliers and can be missing in less developed parts of the world (and may be something that requires investment).
- Developing skills and competencies in a sustainable way that reinforces the use of local education providers while meeting international standards can be challenging in places where the education system (which we address in the next section of this book) is substandard.
- It can be difficult, particularly in the current economic climate, to find the right funding sources to sustainably support the growth of scalability of local companies as well as to support their acquisition of technology and tools that are fit for purpose.
- Financing small-scale suppliers is always difficult. A lack of finances and the general state of the economy works to keep local suppliers limited in size and confined by market restraints (a lack of free-flowing capital to finance small companies, etc.). This means that local suppliers are often unable to improve their capacity to supply through growing their business (either in terms of skills training, hiring new employees, or buying the right equipment) and are consequently incapable of meeting the demands of the oil and gas industry.
- Multinational companies with experience of helping local supply chains grow and develop often report that the attitude of local individuals is a significant challenge. Local suppliers can be relaxed about meeting deadlines and achieving high-quality results. Often this is a result of the context of the local economy, where work is sparse or intermittent, and there is sometimes a lack of emphasis on meeting deadlines and communicating potential delays. One might read this challenge as a 'clash of working cultures'. That said, this may also illustrate a certain negative bias from international operators towards local suppliers.
- Building the capacity of local suppliers hinges on return on investment (ROI). There is a tipping point at which investment in local supplier development becomes unprofitable for operating companies, and this fact must be understood. Too little or too much investment in local supplier development can lead to a loss.

- In many regions of the world, both governments and multinational companies lack a detailed and holistic understanding of what skills are required to improve the capability of the supply chain. This can lead to misaligned efforts and wasted investments. Operating companies and governments should focus on building nonspecialised skills among their technical workers, not training geoscientists and petroleum engineers who may not actually be needed.
- In addition to developing nonspecialised technical workers, operating companies and governments must also look at life-long development, providing opportunities for talented individuals to become specialised technical workers with high-level engineering and scientific ability. IOCs can sometimes overlook this aspect of supply chain development (developing the supply chain as the hydrocarbon sector develops), and this is a challenge to be overcome.

2.4.8 Supply Chain: Concluding Remarks

In this section we have briefly covered the importance of supply chain development and why this is integral to any meaningful localisation strategy. Most importantly, supply chain development must be in line with the needs of the oil and gas industry, seeking to identify areas where significant savings can be found by shortening the international supply chain and drawing on local expertise. In practical terms, however, this is easier said than done.

In many instances, international oil and gas operators will find that essential areas of the economy are underdeveloped, such as manufacturing and fabrication. The methods we advocate here are not about oil and gas companies engaging in national development, but rather supporting smaller local vendors, adapting procurement procedures and contracts to enable their participation, and supporting these vendors to deliver to the standards required by the industry. Over time, IOCs and international operators should expect to see their savings at least equal their investments in building local vendors' capacity, but ideally the savings accrued from supporting the supply chain should exceed the initial investments made.

2.5 EDUCATION AND TRAINING

2.5.1 Introduction

The role of education and training in realising any localisation strategy is critical. As we have demonstrated over the course of the previous three volumes of the Getenergy Guides, it is in understanding the dynamic between energy, education and economy that we are then able to reach positive outcomes for companies, for governments and for citizens. In our model of localisation, education and training play a pivotal role in two ways:

- First, to enable local people to take advantage of the direct and indirect (supply chain) employment opportunities that a thriving oil and gas sector generates, governments, oil companies and others have to invest in – and have a clear strategy for – the education, training and development of those local people. In a highly technical, evolving industry like oil and gas, the skills and competencies of the workforce – and their ability to be educated and trained to the satisfaction of employers – is a critical driver for successfully achieving any level of localisation.

- Second, the concept of localisation – of 'doing it locally' – extends to the education and training itself. It is no longer enough to fly people out to existing education hubs like Aberdeen, Texas or Calgary to build skills. Not only is this increasingly unviable from a cost perspective but it fails to localise the education and training provision itself. If that provision is built in country, then access to that provision becomes transformed. What's more, the building of education and training capacity creates significant wider benefits for society, for the economy and for citizens (something we explore further in Chapter 3).

With this in mind, our model of localisation has a localised education and training system that is fit for purpose at its heart. It's important to recognise that there are already many examples of how operating companies and host governments have invested in local education and training provision in the hope of meeting the skills requirements of the oil and gas sector. Many of the best examples of this have been included in our previous books. However, the challenge is often that these efforts are delivered in isolation – an oil company-sponsored technical training centre or a faculty support programme at a university – and whilst such efforts are to be applauded, it is our belief that it is only through a strategic approach to education and training that genuine progress can be made towards localisation. There is also evidence that too often the distance between the provision of education and training for oil and gas operations and the employers for whom that provision is designed to support is too great. Simply having a technical training programme in place – or running a petroleum engineering course at a university – gives no guarantee that graduates from those programmes will be employed or employable. The demands of international companies in terms of technical capability, knowledge and employability are high. Companies want candidates who are 'field-ready' and, too often, local education and training provision fails to meet this expectation.

In our model of localisation, we have amalgamated all of the lessons from our research over the course of the four Getenergy Guides – and our experience of working with host governments, oil companies and educators – to produce an approach to education and training that will enable hydrocarbon-producing economies to maximise the opportunities that exist for their own citizens from the workforce demands generated by oil and gas operations. This model is based on a realistic assessment of what an employable oil and gas worker actually looks like. And, finally, the model takes account of the longer-term legacy of building education and training capacity in country. Before we look at the model we are proposing, it may be useful to understand the challenges that exist around the current approach.

2.5.2 Education and Training: The Current Model

As we discussed briefly above, there are significant challenges to the way in which education and training provision is currently delivered when we think of this as a means of localisation. In many countries that produce oil and gas, the stipulations of local content policy mean that IOCs have a contractual or assumed responsibility to educate, train and develop local people. In these cases, the current model of education and training in many parts of the world is characterised by the following:

- **A lack of shared responsibility between host nation governments and international oil companies**
  International oil companies have often faced situations in developing regions of the world where the employment of locals, and therefore their training and education, is seen as the responsibility of the IOC(s) alone. In Iraq, this has for many years been an underlying factor behind why many IOCs have failed to localise their workforces and why Iraq is still reliant on international workers. IOCs have reportedly felt that the government of Iraq has unfairly placed responsibility on industry without providing the requisite level of local support to make training locally a viable option.
- **Failure to build relationships and trust between industry and education at a local level**
  Local education providers, including technical colleges and universities, have often felt that they have received little interest, investment or assistance from IOCs typically because these providers are not considered to be 'fit for purpose'.
- **A distinct lack of trust between IOCs and host governments**
  As an extension to the previous point, the lack of trust significantly hinders localisation efforts because very little is done to strengthen local education and training systems and build the required infrastructure to create a steady pipeline of local workers. This means that educational programmes, even once established, encounter difficulties as success relies on collaboration between education, government and industry, and this is often simply not there.
- **A preference for in-house or trusted training among IOCs**
  This can lead to a neglect of more sustainable measures to support a localised direct and indirect workforce. Often, after the recruitment of new people into the organisation, IOCs will administer a period of training to the individual, whereby they transition from a trainee to a contributor to the business. Some (and in many cases most) of the training is done outside the country, sometimes using third party training providers. This approach can work well for building the direct workforce, but does little to support the development of local training capacity. In addition to these issues, by not investing in the local education infrastructure, the national education system does not receive the opportunity to grow its expertise in oil and gas and associated disciplines.

- **Lack of co-ordination across the industry**
  The in-house approach we defined above also hints at a further problem with the current model. Although there are exceptions, in too many cases oil companies take an isolationist approach to education and training. This means that they are reluctant to engage in any meaningful way with others from across the sector. As a consequence, it makes it very difficult for governments and local providers to understand what the industry wants or needs as a whole and to then design an education and training system that is genuinely fit for purpose. In this case the local system will always be second-guessing the industry.
- **A preference for international operators**
  As with their approach to recruitment, many international companies fall back on tried and tested models of operation when they think about educating and training their workforce. This means both a reliance on in-house training (see above) and the use of international training providers, usually in locations that are far from the sites of oil and gas operations. Again, this approach is both costly and fails to make any sort of contribution to the local capacity of education and training institutions.
- **Piecemeal investment**
  The development of any education and training systems relies on a strategic approach that works over the long term and an approach that is fully funded at every stage. In too many cases, developing countries fail to implement such a strategy and fail to recognise the sustained investments required. Furthermore, the investments that are made by international companies are limited to small, targeted projects, and the life cycle of these projects is often hindered by a lack of long-term funding. Meaningful change in any education system takes time, and therefore this approach can be enormously counter-productive.
- **To much focus on higher education, not enough on technical and vocational**
  Where investments in local education and training capacity are made – particularly by IOCs and other international companies – there has often been a propensity for these investments to be focused on the higher education system. Although these investments can make a difference, they are often chosen because they are the easiest to make, or because there is an established assumption that the higher level qualifications will bring greater public kudos and recognition from government, rather than because they will have the greatest impact on real skills needs. Local universities are simple to identify and agreements can be reached for supporting staff development and research programmes and so on. Much more challenging, however, is to have a real and lasting impact on the technical and vocational system. This requires more sustained efforts, thinking and, most importantly, money. Without such investments, the TVET system will not benefit. And with the majority of both direct and indirect job roles needed technical or vocational training, any vision of localisation is unlikely to ever become a reality.

Having examined the current model for education and workforce development, it is instructive to look at how host nation governments have typically sought to support these efforts. Below is a list of key points related to this current model. It must be stressed that while these are common features found in various parts of the world, they are not universally present in every case:

- **The establishment of sovereign wealth funds**
  Sovereign wealth funds are a common mechanism used by host nation governments to manage the funds available for investment into projects that can contribute to national development goals. However, these wealth funds, commonly referred to as National Resource Funds, typically struggle to produce the level of workforce development required. This is often attributed to mismanagement or the lack of a coherent strategy for investment. In Iraq, IOCs have, for a number of years, contributed to a national resource fund which is designed to support education and training investments. Anecdotally, very little has been done by the national government to leverage this fund and invest in education and training capacity at a local level (demonstrating very clearly that the challenge is not always one of funding). This has led to considerable distrust between the IOCs, who continue to support local education through other efforts but feel as though their contributions to the education fund are not being utilised for the intended (or at least stated) purpose. Iraq's national resource fund is estimated to be in the region of half a billon dollars, and there is little information as to what has been achieved. There are examples of similar funds that have greater impact but these funds tend only to work if accompanied by a coherent localisation and investment strategy.

- **Lack of connection to the ecosystem**
  IOCs have and continue to build training centres, where nationals can receive the education and training they need to be suitable for employment in the industry. However, while this is laudable, it should be noted that the building of training centres requires detailed planning. These training centres must become hubs of activity, linked to local and international institutes. Without this ecosystem approach – which must be led by government – training centres are likely to fail. Host nation governments – who may support the construction of an IOC training centre – generally do not help in the planning or execution of the project or in understanding its potential legacy. Without the ongoing interest and support of the government, these training centres will not become embedded into the wider education and training eco-system, and they will fail to build the links to other stakeholders required for real and sustained impact. If the host nation government neglects this duty, IOC-funded training centres will simply serve the inherently narrow interests of the IOC. The fatal flaw is often to name the training centre after the company whose funding has

made its construction possible – this mainly serves simply to increase the likelihood that the centre will not be supported by other companies and ultimately will prove to be unsustainable.

- **No plan or strategy**

  Host nation governments in many instances can be criticised for a lack of detailed, informed strategic planning. In a majority of developing countries, the education sector is not given the attention and investment required to bring nationals up to the levels of employability which companies seek from their counterparts in developed nations. In many host nations, governments produce national development plans that cover a specific time frame (such as Ghana's 'Vision 2020' plan). These plans often lack the detailed and practically driven framework needed to achieve their goals. This brings us to the next point.

- **Lack of understanding around demand**

  When designing a plan to develop the workforce for oil and gas operations, host nation governments often do not grasp the complexities of workforce development for a nascent oil and gas sector. There is typically a misunderstanding of what skills will be required across the life cycle of operations (something we explore across these pages) and where the specific demand for workers exists. This misunderstanding then leads to a failure to map skills needs to a skills provision plan.

- **Lack of clarity around supply**

  There is often a lack of knowledge around the capacity of the local education and training system to meet skills and workforce development needs. This relates to poor knowledge of the capacity and quality of local institutions and little coherence at national level around the pipeline of talent that will emerge from further and higher education over forthcoming years. Without fully understanding the ability of the education and training system to supply suitably qualified people – and in the right numbers – it is impossible to design a coherent education and training strategy.

- **Failures of partnership and collaboration**

  Generally, we observe a failure to understand how international education and training providers can be utilised and how to form partnerships between international providers and local education providers. Every country needs the insights that their international counterparts can provide. In Kazakhstan, for example, the national education system – at the TVET level – has benefitted immensely from the experience and guidance of SAIT (Southern Alberta Institute of Technology) – a Canadian technical training institute world renowned for its training capabilities. SAIT works as an international partner to Kasipkor, the state-owned holdings firm that is responsible for building Kazakhstan's TVET system (see case study later). However, too often, the concept of localising education and training provision fails to recognise the need for international collaboration and, as a result, the quality of what is developed does not meet the standards required.

- **The inflexibility of the system**
  Host nation governments in developing countries often have a highly centralised approach to education and training, with technical syllabuses being set by the government and giving little room for local institutes to adapt their courses to the needs of various industrial sectors. Such an approach often fails to engage effectively with those industrial sectors, meaning that a true understanding of what is required is never fully developed. This centralised approach to TVET does not adequately support industry needs and falls short of developing nationals to the standards required by IOCs.
- **Poor governance**
  Host nation governments and their related agencies often suffer from poor collaboration, co-operation and communication within their own departments. This means that responsibility for the education sector can be divided up across several ministries/agencies, creating confusion and misalignment. This largely uncoordinated division of responsibilities not only makes improvements to the education sector and the development (and then execution) of government policy almost impossible, but it also creates a confusing environment for IOCs and limits effective communication between host nation governments and IOCs for workforce development.

While these broad trends relate to the characteristics of training and workforce development in the oil and gas sector – and highlight many of the challenges with the current models – there are numerous examples of positive education and training programmes (although many, as we have previously mentioned, are operated in isolation). That said, we do often see a set of commonly occurring negative features relating to the approach that governments (and industry) take to workforce development. Chief amongst these is the lack of strategic planning and a failure to understand (and map) the skill needs of the sector to the provision of skills. Host nation governments must, from the beginning, understand the needs of the industry as a whole (including the extended supply chain) and focus the provision of training that can provide the *needed* talent, not the *assumed* talent. Broadly speaking, when IOCs enter a new area of production, governments – keen to announce the job creation levels – will not manage expectations around how the nascent oil sector will benefit the national economy. Education institutes and the government then set about building capacity in what are essentially niche areas of higher education like geoscience and petroleum engineering. In reality, IOCs will not look in-country to find geologists and geoscientists, and may actually need five people with these skills, not 500. Governments need to understand where the job growth is (including job growth in the extended supply chain) and how they can work with IOCs and other industry stakeholders to meet their skill needs realistically and locally.

Kazakhstan – Building a Local TVET System

Since 2014 Kazakhstan has embarked upon a plan to modernise its TVET system. Kazakhstan, one of the most important producers of oil and gas in the world and the biggest producer in the central Asian region, has suffered from a severe lack of well-trained, competent technical people for its oil and gas sector, upon which the country's economy depends, and this modernisation approach is their response to that challenge.

The lack of technical people for the oil and gas sector stems from two important factors:

- First, in Kazakhstan, the low level of investment into the TVET system has contributed to a national mindset, whereby nationals see vocational and technical education as being subpar to the higher education system (university). Individuals who opt for a technical education usually face stigma for this decision and are viewed as less 'academic', less competent and unable to pursue a higher education at university. This has caused surprising statistics to emerge: official reports show that only 36% of young people choose to pursue a technical education route, while 63% go on to university. This is in spite of the fact that industries such as oil, gas and construction actually need technically trained students, not academic graduates.
- Second the education system in Kazakhstan continued to resemble – until very recently – the outdated Soviet model of education that, though robust and effective in its day, is now unfit for purpose in Kazakhstan's rapidly modernising economy. This system relies on outdated modes of teaching and learning and is overly focussed on academic achievement rather than competency development.

With these two factors in mind, the Kazakh government went about developing a coherent plan to modernise the education and training systems of the country along with its economic modernisation, privatisation and diversification plan, which has been tremendously successful.

Kasipkor, a state-owned company, was introduced to oversee the transformation of the TVET system and ensure that modern facilities were created to compliment a strong vocational and technical education system. This has meant forging international partnerships with education providers from areas of the world where the education and training systems have successfully developed a workforce that can handle both conventional and unconventional oil and gas production.

The Southern Alberta Institute of Technology (SAIT) was chosen as the international partner and, since 2014, the Canadian technical college has worked with Kasipkor to develop its oil and gas training offer, focussing on centres of excellence throughout the region. This collaboration has resulted in the establishment of APEC, the Atyrau Petroleum Education Centre. The facilities house state-of-the-art education technologies designed for the development of oil and gas competencies. The Centre was launched delivering three key courses, but in the run up to 2020 will provide more than 17. These include the following:

- Petroleum engineering technology
- Production field operations
- Instrument technician
- Chemical engineering technology

Continued

Kazakhstan – Building a Local TVET System—cont'd

- Power engineering technology
- Power and process technology
- Gas engineering technology
- Drilling technology
- Instrument engineering technology
- Electrical engineering
- Industrial equipment
- Heavy industrial equipment

Graduates, upon completion of their training, receive two qualifications: a Kazakh certificate providing local recognition of the qualified status of the candidate as well as an internationally recognised certification from SAIT.

Kasipkor relied on SAIT and the college's experts to conduct research into what sorts of training equipment would be most appropriate for APEC. SAIT researched the process equipment and various engineering technologies currently in use across Kazakhstan and then selected training equipment to mirror the actual skills required in the region. Using this research, Kasipkor developed a list of training equipment and then asked for the industry's feedback. Once approved, Kasipkor then purchased the training equipment for installation at APEC.

To ensure that APEC becomes self-sustaining and that Kazakh teachers and trainers are part of the solution, Kazakhs are being trained by SAIT staff, giving them the opportunity to update not only their technical skills, but also to receive pedagogical coaching from some of the world's best teachers in technical subjects. The approach will involve phasing out the role of SAIT teachers over time and transitioning to a point where APEC is entirely staffed by capable Kazakh trainers.

The Kasipkor model, whereby the government established a company to oversee the rejuvenation of the TVET system, international education experts were brought in to partner on the development of a new state-of-the-art training centre, and the industry approved the syllabus and training equipment, presents one of the strongest and most sustainable approaches to meeting the challenge of training nationals for oil and gas operations (Information taken from The Getenergy Guides, Volume 2).

2.5.3 Education and Training: Relevant Education and Training Trends

There are now numerous emerging trends for education and training in the oil and gas sector that are making local education and provision a much more viable proposition across different parts of the world. A key factor here is the increasing use of technology and the impact this has on the portability of education and training provision. The following trends are now shaping the way that oil and gas workers in the 21st century are trained and educated:

- **Simulations and the use of virtual and augmented reality training tools**
  Simulation training provides high-risk industries (those industries that carry inherent physical risk factors to staff and the wider environment) with a

risk-free context in which key competencies can be developed. The recent evolution of virtual and augmented reality systems – including considerable reductions in cost – have made these compelling training technologies available to a much wider audience.

- **E-learning programmes and tools**

  This is still a huge area of development for the oil and gas industry, and many companies have been slow to grasp the importance of training people using a full suite of e-learning tools. Other sectors have been quick to recognise how effective e-learning can be and the impact that e-learning approaches can have on improving the competency of workers in a very short space of time. In addition to the competency development benefits, e-learning reduces the time spent away from the field and brings a wealth of resources to the fingertips of learners regardless of their geographical location. The importance of e-learning is also connected to the increasing importance of competency-based training.

- **Competency-based training**

  Competency-based training, or CBT, is designed to provide real, applicable skills through teaching those skills in a practical, hands-on setting. The idea is that you 'teach through doing', and this then translates into competency and 'field-readiness' to undertake the job roles that need filling.

- **The power of 'theme park' training**

  Theme park training is a form of immersive training whereby students are put into realistic operating conditions, where every move matters and there are real dangers and consequences to the actions taken by candidates. Inside the 'theme park' or operating facility (which might be a decommissioned offshore platform) risk is reduced by the fact that the facility has no real production capabilities. One example of the theme park model in action is the INSTEP facility in Malaysia used by Petronas.

- **Accelerated learning**

  Accelerated learning is still regarded as a controversial approach to providing key technical competencies, yet it is gaining popularity. The accelerated learning process exposes students to the realities of the work environment through maximising the practically orientated components of the training. This allows for core competencies to be attained far quicker than traditional methods of training students and reduces what many IOCs refer to as 'time to competence'. Not only can this approach help develop people faster through training programmes that harness proven practical training methods, it is more financially viable for the following reasons:

  - It reduces the time a new recruit can be turned into a contributor.
  - Training focusses on the most industry-relevant practices and can be aligned to business needs.

- Retraining and up-skilling can happen over shorter periods and reduce the amount of time an individual is away from their core duties.
- The current price of oil is placing significant financial restraints on the training budgets of oil and gas companies. Through shorter, more targeted training, companies can significantly reduce the cost of training.

These trends underpin a growing need to train people faster and in risk-reduced (rather than risk free) environments. These approaches also crucially enable trainees to become contributors in a much shorter time frame. In today's market, training must have an immediately perceivable ROI for companies, and this need is the driving factor behind the use of new training technologies and new approaches to workforce development. For any localised education and training strategy – which is likely to be led by government and local providers – to take hold, these trends and approaches must be taken account of as they are prevalent and are now being adopted across the industry.

The Algerian Petroleum Institute – Accelerated Learning

At the Algerian Petroleum Institute they have adopted an accelerated learning model. This was implemented after a massive recruitment drive. The idea has been to bridge the skill gap through significant recruiting and then accelerated learning that reduces training from 9 to 4 months for some technical workers. Retirees have been engaged as part-time teachers, enlisted on programmes geared towards new oil and gas workers. This is enabling a high level of knowledge transfer to new recruits. Retirees are also utilised as mentors, and this gives students a personalised way of interacting with seasoned professionals, providing significant opportunities for learning. Accelerated learning is about providing students with real training. Simulation training and competency development programmes play a role, but at its core accelerated learning is teaching by doing and it requires having mentors, being supervised, and having a relationship with direct superiors.

2.5.4 Education and Training: What the New Localised Model for Education and Training Should Look Like

Oil and gas companies need to balance training and education with maintaining business operations, meaning that oil and gas workers need to be both well-trained off site and actively engaged at production sites. The challenges are numerous for companies that are seeking to strike this balance.

IOCs can struggle to understand where to make education and training investments and how to get the most from the investments they make in building local capacity. Ultimately, designing and building a training centre, providing technologies to local companies, building labs or donating equipment to a university must all make 'business sense'. Business sense means making an investment in capacity building so that, over time, the skills and talent pipeline provides a steady cadre of new professionals who are ready to be contributors to the business. Any education and training strategy, therefore, must contain certain key elements. In the following bullet points, we highlight some of the principles of the new model of education and training that should emerge if localisation is to be effective and strategic.

- **The skill needs of the industry must be understood intelligently**
  From the outset, when governments start announcing the potential job growth (which often tends to be exaggerated) there is insufficient understanding of the types of skills oil, gas and OFS companies will need across the project life cycle. A strategically driven oil and gas TVET system will, ideally, understand the skill needs over time, mapping the scale, size and makeup of the workforce to meet the needs of the industry. For example, understanding that fewer engineers are needed than technicians and maintenance people offers a significant steer on where investments in education and training can be usefully made. Such an approach means understanding and developing the professional skills required alongside the technical skills – people with accounting and administrative abilities, as well as IT specialists, systems analysts, economists and so on. Such an approach also requires the development of capable public sector workers so that the government can evolve to meet the needs of a growing, stabilising economy. This task connects with the skills and workforce surveys that we suggest in Section 2.2 – it is only by understanding, in detail, exactly who you need, how many people you need, when you need them and at what level that you can design an education and training system that is fit for purpose.
- **The approach needs to be systemic, not individualistic**
  IOCs must seek to not only build training centres (and this should only be done where it makes solid business sense to do so), but also seek to create ecosystems around existing institutes of higher education. This networked approach – which should take into account every level of education relevant to oil and gas operations and the wider supply chain – can ensure that there is genuine long-term value created from investments. In Houston, the Community College Petrochemical Initiative (CCPI) has significantly helped to address skill deficits in the area related to oil and gas work (see CCPI case study) and is an example of what can be achieved if local partners work together and work in tandem with industry.

Community College Petrochemical Initiative (CCPI)

Houston's community college system can take credit for helping to make Houston the energy capital of the world. The community college system in Houston sprang up around the state's nascent oil and gas industry and, from its earliest days, had to be responsive to the oil and gas industry, upon which the state's economy was reliant.

The CCPI started in 2013 when ExxonMobil engaged Lee College, a prominent technical higher education institute in Houston, Texas. ExxonMobil was in the process of expanding its local operations significantly, estimating a US$12–$14 billion investment for some 22,000 additional workers. ExxonMobil knew it would need to up-skill its existing workforce while at the same time embarking on a recruitment drive. However, given the deficit of technical people in the United States, ExxonMobil wanted to create an ecosystem around local universities and colleges that enjoy a good reputation for the quality of their technical education and training.

In 2014 the number of colleges involved in the CCPI totalled nine with all the colleges situated along the east coast of Texas. ExxonMobil knew that the workforce crisis was so acute that it would take more than just an MoU (memorandum of understanding) signed with one or two universities. Any initiative would require collaboration, not just between the industry and the education sector, but between individual institutes within the education sector as well.

Initially a steering group was established to ensure that each college could communicate with each other and with ExxonMobil in a forum where sharing could take place and strategic development of the initiative could be discussed.

To date, the CCPI has been successful. While hard to gauge in exact terms how many new recruits the initiative has helped to place within the industry (since the initiative only started in 2013) its long-term success is highly probable. The initiative has helped to raise awareness of the skill gap, attract students into the industry and create the data sharing systems required to analyse the skill pipeline for the industry as projects move forward.

- **Communication is key**

  Ongoing communication between industry and the education sector is essential. This is best structured within a tripartite approach, whereby the government, industry and education work together. This helps to build a responsive education sector, create the education policy required to upscale the education system as a whole in light of investments made by IOCs, and IOCs can then create effective ecosystems that operate in partnership with technical colleges and universities. This should help to ensure courses remain relevant to the industry, promote relevant practical experience and create opportunities for industry placements. Additionally, it will also help to ensure that the mechanisms of government are established and the skill needs of the industry can be tracked.

- **Labour market information underpins any system**

  The need for detailed and ongoing reporting on labour market data (and effective dissemination of this data to interested parties) is vital. This entails continuous and properly conducted market surveys involving the right companies and people and focussing on the breadth of direct and indirect employment areas across the economy. Part of the challenge is to establish effective channels through which data can be gathered and disseminated. The UK offers a powerful model for the collection and sharing of labour market information in the form of Sector Skills Councils. Sector Skills Councils track the demand for skills and watch for industry changes that impact the education and training requirements for people working in a given industry. This may relate to the number of people required, the type of people required or the specific skills that such people will need. This information is then fed back to education providers and into education policy so that the system – and the institutions within it – can be responsive to these needs.

- **A stable relationship between the IOCs and the government is key**
  Often governments and multinational oil companies disagree over where the responsibility lies for investments into education and capacity building. However, investment and cost sharing is only one small element of why shared responsibility is hard to achieve. In situations where there is a readily available cash flow – either from oil and gas revenues or developmental aid – often organisation and co-ordination at the ministerial and governmental levels is lacking, money is channelled into the wrong areas, a lack of accountability and independent monitoring can lead to misconduct, and a lack of communication structures can create misalignment. This diminishes belief in the mechanisms of state. Conversely, if governments can find a way of engaging honestly and positively with industry, the rewards are significant. Trust again is crucial to this process.

- **Properly mapping the education and training landscape is crucial**
  A thorough analysis of the local education landscape not only helps to understand the skill pipeline within the workforce, it can help identify areas of misalignment where perhaps local institutions may be failing to offer the right kind of programmes that industry needs. It is only through understanding the current education and training capacity – and being honest about areas of need – that we can confidently create a strategy for localisation. The approach should always be that, wherever possible, we build on what is already there. Once these areas of need are identified, investments can be strategically targeted based on improving existing provision and meeting skills demands set out in the labour market information studies carried out. This mapping exercise will also help to identify areas where international partnerships can be most usefully implemented.

- **Engaging industry is about incentives as well as outcomes**
  Sharing the responsibility for education and the development of nationals in emerging economies can be fraught with difficulties. Host nation governments will often put significant responsibility on the IOCs for providing the necessary funding. However, it must be stressed that IOCs are not in the business of 'state building', and their investments in education and training must be limited to the scope of their operations and to improving their local business outcomes. However, through incentivising IOCs (through tax arrangements, for example) investments in the development of people

for the oil and gas industry can trickle down into the wider education sector. With the right support from government, and the right policies to build the education system, emerging economies can leverage industry funding as a means of developing education infrastructure and systems.

- **Steer IOCs towards a local solution**
  IOCs must transition from their traditional reliance on international training providers (who usually have a limited and transitory presence in the host nation or are located overseas) to a model that is locally driven and therefore allows greater access to oil and gas training for nationals. This means creating the ecosystems mentioned previously through bringing international training providers into close contact with local providers and ensuring that local provision is able to generate field-ready graduates.

It is evident from the model we are proposing that localisation will entail significant and sustained work and investment in the local education and training system. In essence, this is about strengthening local education and training systems and localising the education process as much as it is about ensuring that employment opportunities are met by skilled local people.

As we can see from the above, there are significant benefits from approaching education and training from the tripartite model, where government, IOCs/OFS companies and the education providers come together and establish the systems and activities around which workforce development can flourish. However, there is a fourth party that can be very helpful to this process - nongovernment organisations (NGOs). NGOs can play a significant role in achieving the vision of localisation, and the role of these organisations is primarily within the education field. The example of VSO is instructive within this context.

VSO and East Africa

Mozambique and Tanzania have been the focus of large oil and gas discoveries in recent years. Both countries are seeking to harness natural gas production to benefit their citizens and accelerate economic development. The size of these discoveries and remoteness of their location has raised questions about how quickly production can begin. The IEA (International Energy Agency) has said that the discoveries 'provide a 75 billion cubic metre (bcm) boost to annual regional output (which reaches 230 bcm in total) by 2040, with projects in Mozambique larger in scale and earlier in realisation'. Exporting LNG will be made relatively simple due to the fact that Asia, a large consumer of LNG, is in direct shipping proximity to Africa's East Coast. China's oil companies have already established their presence in the region and are set to reap the benefits from an estimated $150 billion in revenue to 2040.

However, the large-scale commercial gas discoveries raise numerous questions about the readiness of local people in the region to take up oil and gas work. East Africa has a significant informal workforce, with technical skills being found primarily in this area. This means that apart from having no official qualifications, technical people with skills in welding, pipefitting, construction, etc., are largely trained through unofficial and informal practices (father to son skills transfer, as one

VSO and East Africa—cont'd

example). While local people do have skills that can make them readily deployable in the oil and gas sector, international operators need to ensure, first, that they are meeting international standards of HSE – which is nearly impossible given the unofficial and informal nature of these workers – and, second, that they have some way of referencing workers' skills and capacities for the purposes of quality assurance.

VSO, a leading international development organisation, sends volunteers with significant international experience to assist in development-related challenges. Most recently VSO has been engaged with vocational training strategies in Tanzania so that local workers can receive the important educational and training interventions required to move them from the informal economy into formal employment within the burgeoning oil and gas industry.

VSO works with partners on the ground to ensure that local workers have access to their natural resources via work opportunities. This is done chiefly through education and training. VSO continues to work with the government of Tanzania to strengthen the local education and training systems and infrastructure. Additionally, VSO has helped Tanzania realise policies that create a more stable education and training landscape. VSO, through the Secure Livelihood Programme, has managed to achieve the following:

- Raising the income of poor and marginalised people and helping to support communities through sustainable natural resource management. This has been done primarily through VSO supporting vocational and technical education in country.
- In Tanzania, VSO is working with major international energy companies, including BG/Shell, Statoil, Ophir and Pavilion. Additionally VSO is working with the government to develop vocational skills in Lindi and Mtwara to enhance the quality of education, and this has been happening since 2014 and represents a £1.95 million investment over a 2-year period.
- The programme has been successful. At the end of the first phase 51% of graduates had jobs within 6 months, and VSO hopes that the figure will reach 95% before the end of the programme.
- VSO has introduced City & Guilds standards, using these as the accreditation model for the vocational and technical education in-country.
- VSO is working with VETA, The Vocational Educational and Training Authority of Tanzania, in curriculum development in areas like welding and mechanics. In the future this will include rigging and scaffolding. Later, this programme will move towards overseeing the oil fund so that funding is channelled into specialised training in line with the needs of local supply chain vendors who are supporting the emergent national oil and gas industry.
- VSO secures 3-month-long internships for students in the private sector, giving candidates the opportunity to understand the realities of the industry and gain practical insights into different aspects of the job.
- VSO has also secured high levels of female participation, with one-third of students being women.
- Skill readiness training has been integral to the process so far. VSO set up language labs to improve the English of locals as well as build on their nontechnical skills.
- VSO ensures quality through the Industry Linked Advisory Committee (ILAC) who supervises curriculum development and continuous improvement of training.

2.5.5 Education and Training: Constructing the Model Education System for Localised Skills and Workforce Development

2.5.5.1 The Model for TVET

So far we have looked at the various components of an education system and at how governments, the oil and gas industry, nongovernmental organisations and the education sector can work together to enhance local education and training systems. Although the exact process by which this model is implemented will change and adapt according to the needs and geography of each country, there are some clear characteristics that we can highlight beyond those which we have already covered.

Key characteristics of the model education system include the following:

- In any system of education, having the right trainers/teachers is essential. In technical training the right teachers have an understanding of the industry (in this case the oil and gas industry), a hands-on knowledge of working in this industry and combine this experience with pedagogical ability. This involves having train the trainer programmes in place to regularly update the knowledge of trainers and keep them in touch with the evolution of the industry.

- Having the right facilities to house the required equipment for technical training is very important. As we have referenced already, the need to conduct accurate market research is vital if the TVET system is going to be responsive in meeting the skill requirements of local and international employers (see the Kasipkor, Kazakhstan case study as an example). This necessitates the right sort of funding, and there should be a clear sense of shared responsibility between the IOC(s) and the government here. IOCs should invest in centres where there is a business interest to do so, or even establish their own training centres in collaboration with local institutes. However, the government must ensure that money is being channelled into the education sector and used appropriately.

- Obtaining the right equipment goes further and extends into procuring the correct technologies. This can include everything from 3D simulation technologies for training to ensuring that IT suites and Internet access is provided to all students. This will involve a clear audit of existing technologies available and used across local institutions and an analysis of need that relates to the required technologies across different disciplines.

- At the heart of any effective TVET system are the programmes of learning that students undertake. These programmes need to be competency based, in line with industry requirements and updated on a regular basis. This is particularly true of programmes that relate to the more technical disciplines within oil and gas operations although is also relevant to many other related areas (like accounting, IT, HR, recruitment, etc.).

- Access to the field is integral to the quality of training. Students must be given the opportunity to put their skills to use in the field, work with professionals with experience, and have the opportunity to be mentored by oil and gas professionals. This not only creates better, 'field-ready' students in the long term, but in the short term works as a filter to catch students who may not have the aptitude for field work.
- To build a trusted TVET and higher education system, quality assurance standards must be gradually introduced. This cannot happen overnight. It requires an incremental approach, international partnerships and constant dialogue with IOCs and the industry to support course and programme development. However, engaging in this dialogue creates trust and develops the essential systems and mechanisms required to make an effective TVET and higher education system.

Field-Ready Case Study

The Getenergy Field-Ready programme has localisation at its heart. At the time of publication, the first Field-Ready cohort of students is preparing to start their programme at Takoradi Polytechnic in Ghana. The programme is intended to last for 8 months using accelerated and immersive learning techniques. In this short case study, we examine some of the key components of Field-Ready in the context of a wider localisation agenda.

Field-Ready Is a Simple Definition of the Personal and Professional Qualities Which Companies Seek in New Employees

- We have dispensed with complex and competing standards, levels and sublevels which underpin many formally certified courses and programmes.
- Instead we have worked closely with a number of companies active in Ghana to define the qualities a Field-Ready (i.e., immediately employable) person would exhibit.
- They came up with three specific categories:
  - Technical skills and practical disciplines
  - Applied personal skills
  - Field experience
- Despite spending much of our research time working with technical and operations managers from international service, contracting and operating companies, much of their focus was on applied personal skills rather than, as we had expected, the technical discipline skills. Despite this, to be Field-Ready a person needs to have all three of these elements which make up the curriculum.

Field-Ready Utilises and Grows Within the State Education System

- The programme has been designed to be operable within the state system. To achieve this, it must necessarily incorporate a faculty development programme, taking existing faculty and turning them into experienced trainers. This is a programme which runs alongside the first two cohorts of students going through the Field Ready programme.

Continued

Field-Ready Case Study—cont'd

- This is the principal role for the international oil/gas industry training providers. Their role is to support the college or polytechnic in delivering the programme – by which we mean delivering the faculty development programme and supporting the training. We have incentivised the training provider to remove themselves from the process as soon as is practicable.
- Using a very specific selection system to identify third year engineering students to graduate onto the programme, Field-Ready seeks to overturn the assumption by international companies that the state education system is incapable of producing employable graduates to meet their hiring needs.

Field-Ready Is Determinedly Not a Qualification or Certification

- The key element and differentiation factor for the Field-Ready programme is the system it uses to give successful students their Field-Ready status. The critical aspect here is the companies who are supporting the programme. Under their contract with Field-Ready they provide Technical Commissioners to work alongside the programme, mentor a small group of students and arrange the field experience component of the programme – the one area which colleges and polytechnics cannot achieve on their own.
- The technical commissioners ultimately decide on the basis of their experience whether each student is Field-Ready to their satisfaction. So the Field-Ready programme operates as a field assurance system for new employees rather than a formal qualification or certificate. It intends to be a powerful employment reference that leads to successful Field-Ready graduates being known to and hired into the supporting companies.
- On hiring Field-Ready people the company pays a recruitment fee to Field-Ready which invests some of the revenue in scholarships for the next cohort of students enabling it to select people by merit onto its programmes.

   From our pioneering work at Takoradi Polytechnic, Getenergy intends to expand the programme into colleges and polytechnics around the world and to utilise this enormous network of state education providers in the battle to support meaningful localisation of employment through local, high quality and fit-for-purpose vocational education and training.

2.5.5.2 The Model for Higher Education

In the higher education sphere (as distinct from the technical/vocational sphere), many of the same features are required to create the model system that can provide the oil and gas industry with the people it needs and that can do this locally and at scale. For example, strong policy, collaboration with the IOC(s) and OFS company/companies, and direct funding of institutions are all required. In addition to this market research – including but not limited to labour market information – is required at every stage.

   One caveat, however, is the scale of output that institutes of higher education need to achieve to supply the IOCs, OFS companies and the supply chain with the experts and specialists they need. Generally, the number of high-skilled workers, such as engineers and scientists are considerably lower than the need for mid- and low-skilled workers. Similar to the TVET model, the higher education model requires that, at the pre-workforce development stage, labour market research be conducted to understand the specific numbers of engineers, scientists and geological specialists that will be needed to contribute directly to the oil and gas industry or to fill the supply chain with the talent required to perform such tasks as conduct environmental surveys, consult on maritime activities, shipping and logistics, and for business analysis and economic forecasting. Within this context, there are several key areas to focus on when developing a higher education system to not only supply the oil and gas industry with talent and the wider supply chain, but to also assist in meeting national development goals.

- **Understanding the talent needs in the oil and gas industry**
  To achieve the right outcome and construct the 'talent pipeline' for the future, it is important to understand what areas of the supply chain and workforce will require skills and capacity development, what programmes and courses must be developed and how best to effectively deliver that education to improve capacity in those areas. Again, this requires significant labour market research and ongoing communication with the industry, so that exact figures and requirements can be gauged and the development of those high-skilled people can be undertaken accordingly. It also needs to be recognised that these demands will evolve considerably over time.

- **Curriculum development**
  The curriculum offered at universities needs to address the core technical and nontechnical areas associated with the job roles that graduates will likely find themselves in. If graduates are entirely unprepared for the realities of work, this will impact significantly on the perception of the institute and the employment prospects of graduates from those institutes down the line. Curriculum development requires ongoing engagement with the industry and should see IOCs building education ecosystems around local universities and colleges.

- **Strong relationships and bilateral partnerships**
  Strong relationships and partnerships with relevant international universities who can support programme development and teacher exchanges are critical, and there have been many examples (like that of SAIT which we referenced earlier) that prove this point. The key thing is to develop and keep the talent and knowledge in country and to create relationships with international partners that are mutually beneficial.

- **Attracting, retaining and training the right staff to support the university and its curriculum**
  In part, this can be achieved by encouraging experts from IOCs and OFS companies to teach or lecture at higher education institutes. Many of the specialists in these organisations have strong academic backgrounds

and would be able to communicate their practical experience and knowledge in an academic setting. However, it is also important that an honest appraisal of existing faculty is undertaken and that gaps in knowledge or expertise are addressed. This can be achieved through targeted partnerships with international education providers and through a sustained investment in teacher development programmes. It may also mean recruiting new talent and being open-minded about where that talent is found.

- **Building and investing in research capacity**
  Building the robust research centres that can support and feed into the evolution of the energy industry – and, more generally, building local scientific capacity – is a critical aspect of any long-term plan for an effective, localised higher education system. Not only will this help local institutions develop strong partnerships with industry in the short term, but this will also provide the grounding for both horizontal and vertical diversification away from oil and gas in the long term. Those countries with a successful approach here – like Norway – are already investing heavily in research projects that seek to develop a deeper understanding of future energy sources. That said, many governments have failed to keep their eye on the future and have not utilised oil and gas revenues in the pursuit of research that will drive diversification. Those countries will ultimately struggle as their hydrocarbons run out or, as may happen, demand falls away.

2.5.6 Concluding Remarks

If localisation is the centerpiece of the new upstream oil and gas business model, and if localisation is going to be conducted in a strategic manner to target higher levels of production and lower cost for the business, then the heart of localisation has to be education and training. Oil and gas companies, through strategic investments in education and training, both locally and with international providers, can support the development of their direct workforce and also benefit local suppliers through similar training programmes.

Governments are right to see a burgeoning oil and gas sector as a means to stimulate economic growth and development. However, the key to actualising the potential of a rapidly growing technical sector, such as the oil and gas industry, is to ensure that education and training policy is right, and that the industry–education connection is effectively created and sustained. It is also vital that labour market research is produced in a reliable and systematic way to facilitate proper planning, and guide investments into the right areas of the education and training system. Only through committed and long-term investment in the education and training sector can governments hope to realise the potential of their hydrocarbon sectors and ensure the wealth that accrues is invested in its people.

2.6 GOVERNANCE AND REGULATION

Introduction

In the last section we discussed the role of education in supporting the effective implementation of a localisation strategy, and in that we touched on the role of government and how governments must share the responsibility for workforce development with international companies and others with interests in country.

Governments, in fact, have more responsibility than the IOCs to ensure that the economic potential of an oil and gas industry can be realised for the benefit of the national economy and the development of its people. Poor governance essentially characterises the underpinning drivers for countries that suffer from the 'resource curse'.

The term 'resource curse' is perhaps somewhat misleading. The 'resource curse' lays the blame on the natural resources, in this case hydrocarbons, but the resource curse is only the trigger pulled by a government that lacks the capacity to oversee the rapid expansion of an important and lucrative industry. Governments that fail to properly manage the financial resources that accrue through inward investment into their hydrocarbon sector can cause what is commonly known as capital flight, whereby revenues that should stay in country actually end up flowing out to foreign entities. Furthermore, it is governments who are uniquely placed to create the right type of business environment within which supply chain companies can flourish. And, added to that, it is only governments that can lay the foundations for an education and training system that can meet the challenge of skills and workforce development.

IOCs and other international companies face a particular set of challenges when entering an emerging market and or new oil- and gas-producing country. These are, in very broad terms, the challenges that are typically evident:

- Governments of countries that have little or no prior experience with an oil and gas sector usually do not have the expertise and insights into the industry to be an effective localisation partner.
- Governments can be short term in their thinking, one election to another and often fail to implement the long-term strategies required to achieve localisation goals.
- The capacity of governments to negotiate with oil companies, develop coherent contractual arrangements and provide oversight of the sector can be limited and often needs developing.

In this section, we will seek to answer several important questions, which include the following:

- What are the shortcomings of governments and their policies related to workforce development, local employment and local content?
- How can these shortcomings be overcome to ensure that governments meet the expectations of IOCs when it comes to workforce development?
- How can governments be effective localisation partners?

We will look at some of the ways in which governments can approach workforce development, job creation and supply chain growth through policy creation and through working with the oil and gas industry. Key to this is an understanding of where the government's responsibility lies when it comes to workforce development, what role government policy and institutions can play and how a production relationship between government and industry can be the catalyst for achieving localisation goals.

2.6.1 Governance and Regulation: Weaknesses in the Current Model

While it is impossible to provide an assessment of the actions and approaches of every government in every oil- and gas-producing country in the world, there are some broad characteristics we can apply and some general trends that are commonly observed, particularly where an emerging economy seeks to, or begins, producing commercial levels of oil and gas.

- **Confused oversight of the industry**
  When a developing nation finds oil and gas reserves in commercial quantity, the government finds itself having to manage a new industry that it has no previous experience with. The learning curve is significant for governments trying to oversee a nascent industrial sector, and getting to grips with an industry as complex and globalised as oil and gas in such a short time tends to lead to a lack of clarity around oversight. This can mean blurred lines of responsibility between ministries and regulatory bodies. Often, this leads to a lack of communication between government departments and can have serious repercussions for the oil and gas industry.
- **Lack of stakeholder and cross-sector engagement**
  The lack of cross-sector engagement usually means that the emerging oil and gas sector becomes isolated from the wider economy and treated differently. This has an impact on how the government goes about developing the local workforce for oil and gas training and it affects the strategy around building supply chain capacity. Technical skills and professional skills are not always unique to the oil and gas industry, with many of these skills being transferable and valuable to other areas of the economy. However, the failure to meet basic technical and higher educational requirements in the past and then the rush to develop people for the oil and gas industry can, and often does, preclude cross-sector engagement in wide scale workforce development initiatives. This is tied into the next point.
- **Unrealistic ambitions shared with the nation**
  When new oil and gas discoveries are made, governments are quick to announce that there will be significant job creation as a result. While expectations are usually high around the economic prospects commercial-scale quantities of hydrocarbons bring, governments need to manage these expectations. Usually,

the direct workforce uptake is nowhere near the job growth figures predicted by governments. In fact, the majority of jobs – in almost every case – will be generated in the supply chain that directly feeds the industry. If there is a lack of cross-sector involvement in workforce development initiatives, governments will fall far short of meeting the expectations and job growth figures they predicted.

- **Unrealistic expectations forced on the industry**
  International oil and gas operators often face unrealistic expectations from government. This usually relates to development of local talent, employment rates and the use of local companies. Oil and gas companies cannot, and should not be, primarily responsible for improving the state of the education sector beyond what is reasonable. Instead, this requires government to develop a coherent education policy and ensure that contributions, taxes and revenues coming from the oil and gas sector are invested into education and training provision, and that the policy actually works both for the industry directly and for the wider economy.

- **Mismanagement of funds**
  In many oil-producing countries, we can see how mismanagement of taxes and revenues produced by the industry has a negative impact on the economic and industrial landscape. Sometimes this is as a result of an attempt to prop up an authoritarian state (such as what happened in Iraq, Libya and certain post-Soviet nations) and sometimes this is due to a lack of clear thinking and strategic investment within the country concerned. When the government demonstrates an inability to manage national wealth for the betterment of the economy, international investors are unlikely to view the country as a viable business region.

- **Capacity issues**
  When an oil and gas industry emerges, suddenly you have a government that is required to oversee a major new industry that it has no prior knowledge of. It can be the case that the competence does not exist within government to properly understand how to motivate international operators to invest in workforce development and to partner in other practices that can contribute to localisation.

- **Lack of transparency**
  The lack of transparency – usually the result of all of the factors above (confused government oversight of the industry, multiple ministries, councils, regulatory bodies, unrealistic expectations placed on international operators, a lack of engagement and dialogue with the wider economy) – enables corruption, and this ultimately leads to a version of the resource curse. When enough factors compound, and the misuse of taxes and revenues generated by the industry becomes easier due to the levels of confusion and wilful obfuscation of the government's dealings with the industry, the country suffers.

- **NOC monopolisation of the national oil sector**
  When state-owned oil companies are allowed to dominate the sector and effectively monopolise it through the legislation laid down by government, this reduces competition and can stifle innovation and, ultimately, production capacity. Mexico is a case in point here with the country responding to the problems it has created by reintroducing international players into its oil sector and, notionally at least, creating competition between Pemex and international oil companies for offshore licences. When NOCs are allowed to monopolise the market, they can pave the way for exploitation, corruption and financial mismanagement.

Case: Brazil and Petrobras

At this stage it is important to reiterate that not all developing nations have failed to manage their oil and gas sectors in a way beneficial to the wider economy. Brazil, through the government and the largely state-owned national oil company Petrobras, has successfully managed to create tangible results and localise its oil and gas workforce with some degree of success. Until recently, Petrobras' ability to place well-trained locals in all levels of its workforce (as well as to invest in research around new methods or exploration and production like deep-water/presalt areas) has held the country in good stead in regard of building local content. However, even Brazil, as a broadly positive example of a properly managed oil and gas industry that has grown and helped stimulate the wider economy, has fallen victim to corruption, which has undermined the good work of Petrobras.

This goes to show that, even if the circumstances are right, a lack of transparency and improper management of funds can have a significant impact on the economy. For years, Petrobras was able to forecast its own annual results and plan spending based on those projections to generate international investment. However, the company has failed to meet production targets and has continuously failed to live up to its plans. In 2012, due to huge offshore discoveries, Petrobras announced that by 2015 it would be producing some 4.6 million boepd (barrels of oil equivalent per day). To date the company is only producing 2.8 million boepd despite gaining $250 billion of investment.

This type of mismanagement has cost the company. Petrobras was worth $200 billion, making it one of the top 10 oil and gas operators companies in the world. However, since the collapse of Brazil's economy in the wake of Petrobras' financial mismanagement and ongoing problems of government corruption, Petrobras is now worth $56 billion.

In addition to this, the company has accrued huge amounts of debt, which it will struggle to pay off given its production levels and the faltering market worth of the company in its present state. We can clearly see from the Petrobras example that financial mismanagement and the monopolisation of the NOC over the oil and gas sector may have left the country without the international investment in one of its most important sectors.

2.6.2 Governance and Regulation: Making Governments Effective Localisation Partners

Now that we have highlighted some of the ways in which host nation governments fail to be effective localisation partners, it is important to highlight some of the key objectives for governments as they move forward and convert their hydrocarbon potential into real benefits for their economy, for communities and for local citizens.

Incentivising Investment in Workforce Development

There are numerous ways in which a host nation government can incentivise international operators to invest in human capacity development and support an ongoing transfer of key extraction technologies into the country. While local content policies and legal framework are helpful and, arguably, essential to meeting the priorities of host nation governments, incentivising international investment into a given region and securing inward investment into the local workforce is, perhaps, a more effective strategy.

When trying to create the incentives to firstly attract international operators and secondly facilitate their investment in localisation, it is important to understand the influencing factors for companies, which include:

- Investment policies
- Political stability
- Strong regulatory environment[3]

These factors are commonly listed as most important for investors looking at emerging economies with significant oil and gas production potential. Additionally, governments can provide tax incentives to international companies to incentivise them to invest in workforce development and technology transfer. According to studies conducted in Ghana, oil and gas investors have particularly looked for competitiveness and innovation, and these two areas directly relate to education capacity and to the capability of the local workforce. According to a recent study into Ghana's investment attractiveness:

> the requirement of skilled labor in the oil and gas industry as well as other industries, it is very important for the country to ensure that, its labor force is well trained in the application of high technologies in order for it to be relevant in today's labor market. This is because most FDI [foreign direct investment] is likely to use modern, up to date and state of the art technologies and operate in knowledge driven sectors. Therefore the capability of the country to apply the existing high technologies, rather than producing them is likely to increase the attractiveness of the country for more FDI.[4]

---

3. http://www.iiste.org/Journals/index.php/EJBM/article/viewFile/6655/6799.
4. Ibid.

Developing Local Employment and Education Policies

Local employment and education policies are of particular importance (and we cover the details of this more extensively in Section 2.3). Governments who are able properly to manage and balance the development of people for the oil and gas industry – and, critically, for the extended supply chain – are, in effect, building an education sector that will support the long-term objectives of the country, including the economy within a post-hydrocarbon future.

Local education policies need to ensure the following:

- The creation of sector skill councils or other bodies that sit between industry and education and that facilitate an honest dialogue between the two
- A coherent plan for clear and transparent investment of oil and gas revenues/ taxes into building the capacity of the education and training sector
- An understanding of the skill needs of the oil and gas sector and the intersecting skill needs of other technical industries for the purposes of economic management and skills mapping
- Improvements to education-related measuring so that the government is aware of the quality of graduates, their uptake into industry and gaps in their knowledge
- Improvements to the quality of instruction and involvement of the international education sector to support teacher and trainer development
- The launch of initiatives that can quickly address identified skill gaps in the working-age population in the short term to increase employment
- Long-term investments in education infrastructure

This list is not exhaustive and the specific set of policies will be defined at a country by country level. However, no efforts towards localisation will be effective without concurrent implementation of employment and education policy.

Allocating Funds to Capacity Building Within Government and Beyond

Not only should funds be channelled into the education sector to build its capacity, governments must also focus on building their own institutional capacity to understand the oil and gas sector, particularly if the country expects to maintain a significant production life span. This involves understanding how effective taxation and the collection of tax revenues should be implemented, and this means having a competent revenue authority. The government should also seek to improve its administrative capabilities to improve accountability and transparency.

Countries with low administrative capacity will suffer many setbacks as they seek to build a capable regulatory environment to oversee the sector. Where there is low administrative capacity, governments should engage external technical assistance and commit to improving the capacity of the government across the board, bringing in international development agencies to assist. The government must also recruit and retain skilled staff, and this may involve revising current pay structures to attract the people it needs into civil service.

Improving Accountability

It is important that the government be held accountable for its decisions and oversight of the sector. In the early stages it is essential that a regulatory body be established to oversee the sector and manage relations between the government and the various actors across the sector. This is generally easier said than done. Governments who have rushed to do this have often found themselves needing to create several iterations of this regulatory agency, and this complicates the issue since once established, it can be difficult to wrest control back from the agency.

International operators do not need to see a central, regulatory authority emerge overnight, but instead would like to see the beginnings of this agency established and allowed to grow and assume responsibility in conjunction with the growth of the sector, the needs of the companies, and the need of the government for ever more capable individuals with the right administrative skills to oversee the development of the oil and gas sector.

Managing Public Expectations

Perhaps one of areas of a government's duty that is often overlooked is the need to manage expectations. All too often host nation governments announce huge employment and economic growth figures to the public. These figures become the basis of hope which, almost always, fails to manifest into actual job growth, which ultimately leads to disappointment and public mistrust in the government.

Governments should provide realistic figures and economic data and try to moderate public expectations around oil and gas discoveries. In the early stages, prior to any economic data being officially released and long before job growth projections are made public, host nation governments (and the agency developed to oversee the industry) should seek consultation with the IOCs and OFS companies entering the sector. Host nation governments often have little to no internal knowledge of the oil and gas industry and are thus not ready to make projections related to economic growth and job creation prior to extensive discussions with operators. Once these discussions have been held, a more accurate set of figures can be discussed with all concerned.

2.6.3 Governance and Regulation: Concluding Remarks

In this section we have explored three key questions which include the following:

1. How can governments be effective localisation partners?
2. What are the shortcomings of governments and their policies related to workforce development, local employment and local content?
3. How can these shortcomings be overcome to ensure that governments meet expectations when it comes to workforce development?

To summarise this section, it is perhaps best to address each question, providing a clear answer for each.

First, governments – if they are to be effective localisation partners – must recognise that several things must be done:

- Improve their own capacity to effectively administer and oversee the industry
- Establish a fit-for-purpose agency that can manage government–industry communications
- Understand the skill and workforce needs of the industry
- Establish a body that can moderate between the industry on the one hand and the education sector on the other
- Create education and training policy that is coherent with both local capacity and with the labour market requirements
- Manage public expectations regarding job creation
- Understand that oil and gas companies will react negatively if they feel that they are solely responsible for workforce development
- Manage the revenues from oil and gas operations in a transparent and productive way
- Work with the industry as enthusiastic partners, rather than working above the industry as stern overlords

Second, governments often fail to be effective localisation partners because they lack the institutional administrative capacity, in-depth knowledge and understanding of the industry, and the effective policies to govern industry behaviour. In addition to these shortcomings, many governments in emerging nations have placed unrealistic expectations on oil and gas companies for delivering on workforce development goals. This is unfair and is a symptom of the ongoing administrative incapability of the government.

Third, to overcome these shortcomings and become effective localisation partners, governments should engage the various players in the oil and gas sector and enter into an initial stage of discussion and consultation related to localisation of the workforce and the supply chain, understand the needs of the industry from these perspectives, and devise a long-term strategy to facilitate realistic job growth that meets moderated national development goals while at the same time being economically viable for the companies investing in the sector. This should also be done in conjunction with international experts in the area of localisation/local content building and in partnership with relevant development agencies.

Chapter 3

The Legacy of Education and Training

Chapter Outline

3.1 INTRODUCTION

The approach to localisation that we have outlined in this volume is, we believe, the only effective, sustainable model for the upstream oil and gas industry in the 21st century. Our belief in this model is driven very much by our understanding of the dynamic between energy, education and economy and by a commitment to the idea that it is only by implementing an effective approach to localisation that oil and gas operations can benefit governments, communities and citizens in host countries and remain commercially viable for oil and gas companies in the long term.

The ideas and approaches that we have presented demonstrate how this can be done, how we can understand the drivers of localisation and how all those charged with implementing localisation should go about it. The end result will be an oil and gas industry that is, as far as is possible and desirable, staffed by local people at every level of management and operations and an oil and gas industry that is supported and supplied by a locally owned and run supply chain. However, this alone is not enough.

Countries that are rich in oil and gas today will find, at some point, that these resources run out. Perhaps more pressing, they will inevitably find that demand for – and prices paid for – their oil or gas will fluctuate and that an overreliance on oil and gas as sources of revenue both now and in the long term is unwise. Although this perspective is reached primarily through looking back at history, there is also a political shift taking place today that further adds to the

    93

uncertainty of oil and gas demand in the future. The dialogue around the role of polluting energy sources in meeting our global energy needs is compelling and the move towards low carbon or renewable fuels is gathering momentum.

Although our reliance on oil and gas remains intact for now, there is clearly a shift that is starting to diminish the perceived future demand. We can see evidence of this in the number of exploration projects that have been abandoned over recent years (the activities of oil companies off the coast of Greenland would be a case in point here). Collectively, these factors point to one thing: it is risky for a country to become too dependent on oil and gas, either as a source of direct revenues or as a driver for employment and business growth. In fact, the term 'Dutch disease' – implicating a causal relationship between the expansion of one particular industry and the relative decline in others – was coined by the Economist to describe the industrial and economic impact in the Netherlands following the discovery of the large Groningen natural gas field in 1959. In this case, the gas discovery drove significant investment and growth in the hydrocarbon sector at the expense of other existing industries. In addition to a reduction of investment, the burgeoning gas industry drew away skills and manpower from other industrial sectors. The Dutch disease became seen as an aspect of the 'resource curse', a phenomenon that describes how countries rich in natural resources can end up with less economic growth, less democracy, and worse development outcomes than countries with fewer natural resources. What we learn from the history of those countries who suffered from this curse – most notably Nigeria and Angola, although there are others – is that even in times of plenty where the oil price is high, countries who fail to address corruption, economic diversification and education and skills development will squander the potential benefits of a thriving localised oil and gas industry. When we consider the emerging political narrative surrounding our use of fossil fuels, the need to address the wider implications of the current model of upstream oil and gas production becomes compelling.

Particularly interesting within this context was the climate change agreement reached in Paris at the end of 2015. Observing the resolve and determination of almost 200 countries at the Paris climate change talks to tackle the impending catastrophic changes to the world's climate, we saw the debate about the merits of the science behind climate change predictions firmly pushed into the long grass. At the very core of the arguments between delegations in Paris was the role that fossil fuels play in warming our world. Burning oil and gas produces harmful carbon which, when released into the atmosphere, traps the sun's energy with a resulting gradual (depending on your perspective) increase in the world's average temperature. Small increases can have huge effects. The world's poorest nations are likely to be the worst hit by the short-term effects, but ultimately we will all suffer.

Over the past three decades the arguments about the reasons why joint action by all nations is impossible have been influenced largely by the benefits richer nations have accrued through industrial development, often at the expense of poorer nations. Why, argued India, China and other major emerging world economies, should we forego the luxury of unlimited energy supply that the developed world has previously enjoyed? Climate change is a crisis created by the established industrial economies of the world, not by us, they would say. The problem is that these arguments are as self-serving as the first world has been since the late 1700s. The difference is surely that we did not know then what we clearly know now. The thicker the smog in Beijing, the deeper the floods in Bangladesh and the fiercer the storms in the Gulf of Mexico, the more our politicians are forced to face the realities of climate change and act in a common cause.

Further evidence of the shift in political narrative can be seen in the levels of investment that are now going into renewable energy projects. The figures for 2015 are particularly instructive: global investments in fossil fuels (in terms of new projects) was around US $130 billion in 2015[1]. Compare this to the figure of US $286 billion, the figure for global investments in renewable energy projects[2]. Of this, US $103 billion was raised for projects in China alone, with India, Mexico, South Africa and Uganda all making significant advances[3]. The figures for 2015 are the strongest – in terms of investment in renewables – for 5 years and demonstrate the growing momentum for a shift towards low-carbon alternatives to coal, oil and gas.

We should work on the assumption that the political tides that swelled up at the Climate Change conference in Paris will achieve steady progress in limiting the consumption of fossil fuels around the world. We should accept that the commitments which nations have made to cap and ultimately reduce their use will change economies and will offer the possibility of a postindustrial clean energy revolution. It is clear that it is much harder being green for nations who still rely heavily on the production of fossil fuels to support their own economic development. What was agreed in Paris should have every oil/energy minister in Africa, South America and across the Middle East thinking very hard about how they are going to change the traditional narrative of dependency surrounding large-scale national hydrocarbon production.

So how should resource-dependent nations square an ever falling oil price, the Climate Change agreement and their nations' well-being and development agenda? There are several 'nonstrategies' which are worth ruling out now. Do not wait for the price to increase. Even if prices do rise, with constrained demand, seemingly inexhaustible supply thanks to new technologies, and the strengthening Climate Change agenda, a focus on the price per barrel will only increase your dependency. Do not rely on your OPEC membership or seek accession to OPEC. Like any cartel, it is only strong if there is demand for what it produces

---

1. From Global Trends in Renewable Energy Investment (2016), UNEP.
2. Ibid.
3. Ibid.

and if it acts in unison. The fractures are already public – it is unlikely to survive until 2020 and its ability to influence the agenda as once it did, is melting demonstrably faster than the polar ice caps.

So what can these nations do? The answer is simple, yet it requires a fundamentally different approach by governments, companies and national institutions. Nations must convert their economies away from resource dependency. They must diversify into new industries and new products. They must seek self-sufficiency whilst continuing to exploit global markets (as those markets have, to date, exploited them).

Many of these nations have national development plans predicated on the increasing economic contribution that governments believe new and existing hydrocarbons will make to their coffers. However, the timeframe for these plans (from 5 to 30 years) already appears unrealistic. Instead there needs to be a new dialogue between resource-dependent nations and resource-producing international companies. An outline of the agenda for that discussion is as follows:

- There should be a broad agreement that a falling oil/gas price and reduced consumption of hydrocarbons fundamentally impacts budgets and the surety of the future for the industry for all those involved.
- There needs to be a recognition that tackling this challenge requires both short- and medium-term objectives which involves new forms of collaboration. That collaboration must take account of the time it will take to reposition the country and its producers away from an addiction to hydrocarbons.
- In a persistently low oil price environment, the industry must be operated as efficiently as possible. That means a real and genuine engagement in the training and education of local workers, the buying of local goods and services and investment in retaining locally as much of the resources produced to be provided into domestic markets.
- In the medium term both governments and oil companies need transition towards new, low-carbon energy production to survive. So agreement needs to be reached that this transition must be embraced before fossil fuels are consigned to the history books.

At the executive level, companies and governments need different financial and strategic plans. They may need new leaders, and they will certainly need to deploy their finite resources in new and much more imaginative ways. At the ground level, both companies and governments will need a workforce populated by people with technical capability, personal skills and practical experience, who are capable of meeting these new challenges. An economy, just as a company, will rise or fall on the shoulders of these workers, most of whom will be technical, operational and managerial staff.

So our argument is that the exploitation of existing hydrocarbon resources and the economic return it can generate must be wholly directed at financing this business revolution. This means finding new and innovative ways of strengthening local and national education systems (particularly at the vocational and technical level) to produce people who have transferable technical, personal and experience-based skills. The only institutions with the capacity required are the universities, colleges and polytechnics in the markets concerned.

If companies and governments can agree to collaborate to transition themselves away from resource dependency today, it might just be possible to plan the transition for a world without hydrocarbons tomorrow. Whilst some might console themselves that these changes are evolutionary in the political and corporate calendar, they are certainly revolutionary by the geological clock. And even if we are thinking of a timeframe of up to 20 years, the impact of the changes to economics, politics, commerce and civil society along the way are likely to be powerful enough to seem like a procession of mini-revolutions with implications for every single one of us.

With all this in mind, our model of localisation should not simply be about meeting the skills, workforce and supply chain needs of the oil and gas sector (although it very much is that). It is about creating a legacy that outlives the oil and gas that is in the ground and that ensures hydrocarbon-producing nations are well-placed to ride the inevitable price fluctuations and shifts in demand, thereby protecting the economic stability of the country in the long term. In this section, we will explore the realities of life as a hydrocarbon-producing nation and analyse the way in which a progressive, strategic approach to localisation can mitigate the dangers that a reliance on oil and gas can generate. Specifically, we will look at the following:

- What the typical life cycle of an oil- and gas-producing nation looks like, and how we can learn from the efforts of others in relation to economic diversification (with particular reference to Trinidad and Tobago)
- How a strategic and coherent approach to localisation can create a skill and talent legacy that supports economic and industrial diversification (and what this diversification might look like)
- What the legacy for the education and training system of the producing country might look like if an effective model of localisation is pursued
- What the implications of this are for IOCs, NOCs and oil field service companies

Through this approach, we hope to enlighten our readers as to a possible future for oil and gas producers of every type where fluctuations in price and demand are not seen as the beginning of the end but, instead, are viewed as the stimulus to create new models of operation that support economic progress and diminish the negative effects of changing demand for fossil fuels.

3.2 THE LIFE CYCLE OF AN OIL AND GAS NATION: THE CASE OF TRINIDAD AND TOBAGO

As we outlined in the introduction above, the life span of an oil- and gas-producing nation is inherently limited. Although many producing countries still enjoy significant reserves, we are now seeing some countries coming towards the bottom of their (oil) barrel. Leaving aside for a moment the argument about fluctuating oil demand – and the notion that we may well end up not extracting significant quantities of known reserves of oil or gas from the ground due to a shift towards cleaner forms of energy – it is instructive for us to look at a country that is simply running out of hydrocarbons and therefore has to face the economic, political and social realities that this brings.

Our case in point here is Trinidad and Tobago. Trinidad and Tobago's oil and gas industry has, over recent years, been forced to explore new ways to respond to weaker hydrocarbon prices and falling levels of production. As recently as 2014, the industry accounted for 42% of national GDP, according to the Central Bank of Trinidad and Tobago, highlighting the critical role that hydrocarbons play in the wealth and economic stability of the nation. Alongside the significant hit that the economy has had to bear in light of the fall in crude oil prices during 2015 and 2016, the country has also had to absorb a major decline in gas prices with key indicators suggesting a price decline of more than 65% since February 2015. On top of this, key gas-based exports like ammonia and methanol have fallen significantly in price since the price adjustments of 2015 unfolded.

The response of Trinidad and Tobago to this challenging industrial and economic context – of falling prices and diminishing resources – is instructive and has two key elements:

First, the approach of the government has been to focus on what one might call 'making the most of what we have'. Within the context of wider macroeconomic policy, this has meant a reigning in of public spending. From an oil and gas perspective, the strategy has been to drive towards raising production levels as high as possible and then investing in ways to exploit the existing reserves – even those more 'hard to reach' hydrocarbons – by implementing new technical approaches to exploration and production that may uncover new sources of supply. These approaches have included enhanced oil recovery (EOR) and have focused on secondary recovery from existing fields to the south of Trinidad, alongside the exploitation of heavy oil deposits in the same area (which have been largely undeveloped until now). Furthermore, the government has sought to generate greater interest in the opportunities for oil extraction by offering up new areas of exploration and awarding licences and contracts to exploration companies.

The 'making the most of what you have' approach extends to finding and then exploiting any potential sources of oil or gas hitherto unexploited. In Trinidad and Tobago, there is significant interest in the Loran-Manatee offshore gas field, which is situated on Trinidad's maritime border with Venezuela. Estimates suggest that the field has around 10 trillion cubic feet of gas and after many years of negotiations with their Venezuelan neighbours, an agreement is now within reach over shared rights to the field.

Searching for new fields is one part of the solution but is inherently limited. And whilst making the most of what you have is certainly an understandable strategy, it is also a sticking plaster which will only stop the bleeding for so long. For one thing, the challenge of generating new opportunities – particularly those involving technical approaches to exploration and production like EOR – is that the cost of bringing these projects to fruition is, in the current (and ongoing) price climate, prohibitive. Simply having reserves is no longer enough. Operating companies need the commercial imperative to develop a field and that is becoming increasingly rare, particularly where secondary recovery is the operating context. There are some levers that a government can use and Trinidad and Tobago could see these applied. A study by the Oxford Business Group highlighted the need to stimulate private sector investment within a context of falling government revenues (they projected government investment in oil and gas to fall by around 72% in 2016–17). Stimulated private inward investment can be achieved through two particular means:

- First, the government of Trinidad and Tobago could make some changes to the production sharing agreements they have to make certain fields more profitable for operators. Although this would reduce the share of revenue, it may create a more dynamic market and encourage new entrants.
- Second, the government could change the way that it taxes the revenues from oil and gas companies. At present, the government imposes a supplementary petroleum tax which is paid when the oil price hits US \$50 a barrel. This effectively pushes up the breakeven point for production to above US \$65 a barrel. Again, in the interests of stimulating private investment in the industry, changing the rate at which this tax applies could be effective.

However, even if the government of Trinidad and Tobago were to implement these contractual changes – and let us bear in mind that these are the sort of changes that other governments around the world are considering to incentivise investments, creating greater competition between countries to secure IOC investment – the challenge of diminishing reserves will not ultimately be addressed through this approach. You are effectively delaying the inevitable and that will only take you so far. Fortunately, Trinidad and Tobago is not only pursuing the 'make the most of what you have' strategy. Far more interesting to us is the second element to their plan.

The second element is, in many ways, pure common sense. If you see that a vital industry is gradually diminishing in terms of what it can offer your economy, the natural – and sensible – response to this is to diversify your economy into other commercial and industrial areas. Earlier in this section, we briefly discussed the 'resource curse' and, as part of that, the 'Dutch disease'. A significant part of these two phenomena – where a country rich in natural resources conversely get poorer and achieves less development as a result of those resources – is the failure to diversify. Before we consider the way in which Trinidad and

Tobago has approached this challenge, it may be useful to understand the specific mechanisms that support – or hinder – economic diversification within a resource-rich economy.

The literature on economic diversification suggests a number of key factors that underpin any drive towards this target. Specifically:

- There needs to be concerted action by governments to create the context for economic diversification; this means that the mechanisms of state – measured in terms of 'quality of governance' – must be effective in supporting economic diversification.
- These efforts need to focus primarily on creating the right conditions for a thriving private sector and for attracting inward investment into the country – states with both these elements tend to be much more diversified.
- Within this context, the economic and business environment must be supportive of private sector growth – this means low inflation, effective but nonintrusive regulation and an environment that is open to trade.
- In addition, significant evidence exists about the connection between the knowledge and skills base of a particular country and the ability of that country to achieve diversification (more of which later).

There are also, alongside these drivers, a number of impediments to economic stability, growth and, by extension, diversification. In relation to Trinidad and Tobago, these include limited human capital, high macroeconomic volatility, inadequate development of infrastructure, inadequate access to foreign markets, rising criminality, lack of innovation, corruption and a burdensome bureaucracy[4].

When we look at the case of Trinidad and Tobago, we can see how the government has attempted to address both the opportunities for, and impediments to, economic diversification. And as we explore this idea of economic diversification, we can start to see how our model of localisation – which has, at its heart, notions of building the skills base, creating local education and training quality, supporting local supply chain strength, building government capacity and so on – is entirely in line with achieving diversification.

It is important to recognise that the debate around economic diversification in Trinidad and Tobago is not new and is something that has been discussed since the 1950s. The country has done very well from exploiting its natural resources and can be considered as a high-income economy, with an oil and gas industry that is globally competitive and underpinned by a progressive and effective approach to education and skills development. However, despite an understanding of the inevitable decline in oil and gas revenues, the nonenergy sector has remained relatively underdeveloped, has failed to attract significant inward investment and, as a consequence, is propped up by government support. That said, there are

---

4. Poverty Reduction and Economic Management (PREM) Network, Economic Premise note on the diversification of the T&T economy; Francisco Galrao Carneiro, Rohan Longmore, Marta Riveira Cazorla and Pascal Jaupart; World Bank, 2015.

a number of indicators that suggest that Trinidad and Tobago has managed to gradually shift away from a direct reliance on hydrocarbon revenues:

- To begin with, the country has seen a steady but progressive diversification away from oil production (which has, in fact, been in decline since the 1980s) towards natural gas, which is now the primary industrial activity within the energy sector.
- Furthermore, the country has managed to grow a healthy petrochemicals sector alongside the maintenance of its hydrocarbon activities.
- Trinidad and Tobago is, in fact, the leading global exporter of ammonia and methanol.

Although this diversification is to be embraced and encouraged, it is what one might consider to be 'vertical diversification' – meaning that it is a diversification of industrial activity within the hydrocarbon sector. In their report on the economic diversification underway in Trinidad and Tobago, the Poverty Reduction and Economic Management (PREM) Network of the World Bank noted the need for horizontal diversification alongside this vertical diversification (meaning diversification away from the resource sector entirely).

One of the critical drivers for this horizontal diversification is human and physical capital. Trinidad and Tobago has been enormously successful in developing the means to educate and train its own people to work in the hydrocarbon sector (with the achievements of the University of Trinidad and Tobago notable here) and actually exports oil and gas technicians to other parts of the world. However, in light of the diminishing reserves across the region, there is a need to explore and understand the transferability of these skills to other parts of the economy (something we do later in this section). In the case of Trinidad and Tobago, the successes that the country has achieved in terms of vertical diversification have not yet been matched by achievements in respect of wider horizontal diversification. In a comparison with other, similar, countries, the World Bank attempted to evaluate the key factors that had affected the diversification efforts in Trinidad and Tobago and to identify, as a result, the best strategy for small, resource-rich countries to adopt in pursuit of this economic objective. Amongst the key findings from this study, they concluded the following:

- That the key mechanisms of state – the governance structures and policy – need to be stable, well thought out and forward thinking.
- That it is critical for countries like Trinidad and Tobago to attract inward investment and international capital, and that this is more likely to happen if there are 'new' sectors of the economy to invest in.
- That in terms of attracting that foreign direct investment, it is common in small resource-rich countries for investments to be largely limited to sectors associated with oil, gas and minerals and that efforts must be made to steer that investment into new sectors of the economy.

- That a failure to develop nonenergy sectors can have a dramatic effect on the economic well-being of resource-dependent countries like Trinidad and Tobago, a failure that can be exacerbated by a lack of support for entrepreneurship and for wider skills development activities.
- That access to and availability of local technology infrastructure can be a critical driver towards attracting inward investment.
- That human capital – and the ability to develop human capital – is absolutely essential to both provide the basis for economic diversification and as a means of attracting that critical inward investment.

The conclusions from the World Bank study were that, on reflection, Trinidad and Tobago had done well in terms of diversifying within the energy sector but had failed, thus far, to effectively diversify into new nonenergy economic or industrial areas and that a great deal of effort will be needed to arrest the economic decline that could evolve as the country's natural resources run out.

So what do we learn from the example of Trinidad and Tobago and how is this valuable to our understanding of localisation?

To start with the positives, we can see how a resource-rich economy can use hydrocarbon production (in this case oil) as a starting point for developing a much broader-based energy economy that can take in gas, chemicals and other related sectors. By doing this, such countries can generate greater levels of local employment, create opportunities for inward investment and mitigate the risk that their economy becomes too reliant on one form of revenue. However, we can also see how failure to diversify into other sectors can lead to a fall-off in inward investment, an erosion of economic stability and a diminishing of employment opportunity and business growth. And within this context, one theme emerges again and again: that of human capital, education and skills. The key to developing a broad-based economy that can withstand both the economic slings and arrows of the global commodity markets and that can adapt to diminishing and, ultimately, disappearing natural resources is to have the skills, talents and competencies to build new industries, to generate new jobs and to attract new investments. This is where the legacy of our model of localisation begins to be powerful in the following ways:

- By ensuring that the workforce that is employed across the industry is predominantly local, a country will inevitably be developing a base of skills that can be adapted to other sectors (more of which later).
- By ensuring that the local supply chain is geared up and able to meet the demands of the oil and gas sector, a country will be building capacity within that supply chain, encouraging entrepreneurship, developing expertise and laying the foundations of business resilience.
- By building the capacity of the local education and training system and ensuring that system is connected in a meaningful way to local industrial activity, a country is able to adapt to the demands of new economic sectors and provide the right level of skill development needed for a thriving economy.

- By investing in governance institutions and policies that work and that are effective in how they support growth in the oil and gas industry, a country is improving the knowledge, skill and expertise of its government and civil service in a way that will inevitably support the achievement of wider economic goals.

It is this that is the legacy of our approach to localisation: a resilient, flexible, highly skilled and autonomous economy that is underpinned by a world-class education and training system bought and paid for with oil money. In the following two subsections, we explore what the skill legacy might look like (and how this can support economic diversification) and what the education and training legacy can be in a country that has effectively localised education and training provision. This is intended to further prove why our model of localisation is compelling and also to act as a broad guide for countries who are looking to evolve from oil and gas production and into new sectors of their economy.

3.3 HOW LOCALISATION CAN CREATE A SKILL AND TALENT LEGACY THAT CAN FUEL DIVERSIFICATION

Throughout this book, we have been explaining how we believe that the localisation of oil and gas operations is highly dependent on a host nation being able to create the local skilled workforce that is able to take up positions in the sector and, furthermore, is able to staff and run the required supply chain. Our model has set out ways that this can be both measured and achieved. The immediate benefits of this approach are clear – the oil and gas sector will become a major local employer, offering jobs, opportunities and economic development to regions where oil and gas operations are being undertaken. However, as we have clearly set out in the preceding paragraphs, any responsible government needs to have one eye on the long-term prospects for their hydrocarbon industry and, almost inevitably, this will lead them to the conclusion that there is a need for diversification. If a country has successfully developed the skills and competencies of its people to meet the demands of the oil and gas sector, this will create a pool of talent that is highly capable, well trained and motivated to work. The big question, therefore, is how can this skill and talent – developed and trained primarily for oil and gas operations – be utilised to support the development of other sectors of the economy? The challenge here is to understand the transferability of the sorts of people who have been engaged in oil and gas activities (both directly in the employ of operating companies and indirectly through the supply chain) to other roles. In some cases, this transferability will be direct – meaning that little or no retraining will be required. But in others, there may have to be some level of 'reskilling' to ensure that individuals are competent and capable of taking up new roles.

Before we explore in more detail this issue of transferability, it is important to recognise two things. First, that there is already significant inward and outward mobility of employees from oil and gas to other sectors. In fact, surveys suggest that operating companies see the poaching of their staff from other industries as being a key challenge in meeting their skills and workforce requirements. That said, some roles are more transferable than others. Second, we should point out that understanding transferability – and then creating a model of economic diversification as a consequence – is an inexact science. There are a broad and varied range of factors that will impact on economic diversification (many of which we touch on in this section) and the skill base, whilst critical, is not the only influence on this dynamic. Taking on board those two broad caveats, it is still relevant to think about what a country – and the people of that country – is equipped to do if they have been effectively educated and trained to work in oil and gas so that workforce diversification can at least be pursued with a modicum of intelligence. In our analysis below, we are not going to attempt to map out the transferability of every job role across the oil and gas value chain – that would require another book – but we will explore some of the job areas and think about how these job areas might be adaptable to alternative economic and industrial settings.

When we think about the skill base of an oil and gas workforce, it is broad and varied. There are a number of models we could use to understand the make-up of that workforce. For our purposes, we will define job roles in the following way:

- **Low-skilled direct employment roles:** This would include roughnecks, roustabouts, crane operators, scaffolders, ROV drivers and other similar roles (generally involving working onshore or offshore at production sites); low-skilled roles would typically require some level of training but this would be limited to weeks rather than months or years.
- **Technical/vocational direct employment roles:** This is the largest area of job families and would include roles across drilling, quality and inspection, maintenance, operations, wellhead operations, welding, plumbing, pipefitting, plant operations, refining and so on; these roles would, again, generally involve working at production sites but would require a much higher level of technical training for those undertaking the role.
- **High-skilled direct employment:** This area would generally be staffed by an individual who had received a degree-level education and would encompass all senior engineering roles, all roles relating to geology and geoscience, specialist refining roles, design roles and some senior technician roles.
- **Office/support roles:** The area of office and support roles is, in itself, vast and varied and generally involves roles that are not base at production sites; roles in this area will include accounting and finance, sales and marketing, administration and contracts, commercial, IT and communications, logistics, human resources and recruitment alongside a range of other roles; staff in these roles will be variously qualified according to their specific job role and level.
- **Leadership and management:** one could consider leadership and management as a subset of office and support roles; these roles are broadly similar to leadership/management positions in any large-scale business and would focus on staff management, project management, business strategy

and management, risk management and financial management. These roles would typically be filled by senior staff with considerable experience either in operational oil and gas roles or at senior management level in an alternative business environment.

Having broken down job roles in this way, we can now start to explore some of the transferability pathways that are available to those working in these roles. We will take each job group one by one:

- First, **for low-skilled workers**, those operating in roles across the oil and gas industry will be familiar with (and trained in) basic health and safety standards. They will also be familiar with shift work and will be used to the challenging physical nature of the industry. These workers could transfer very well into low- to mid-skilled positions within shipping, maritime, construction and engineering. In some cases – crane operation or driving for example – the nature of the work would be very similar regardless of the sector and may not require any additional training.
- **For technical and vocational roles**, trained oil and gas employees may be familiar with working in engineering environments, understanding plans and engineering drawings and completing technical construction or maintenance jobs to clearly defined standards and following agreed processes and protocols. In some cases – welding, plumbing, pipefitting and so on – the transferability is very clear and would generally involve transfer into industrial sectors like construction and engineering. For other roles – drilling, pipefitting, refining, etc. – the transferability may be to sectors such as shipping and the chemicals industry as well as construction and engineering roles.
- **For high-skilled roles,** the transferability is rather more mapped out. For those who have been trained in engineering roles – particularly petroleum engineering – there will be opportunities to transfer into other fields of engineering including mechanical engineering, civil engineering, chemical engineering, and systems engineering. For geologists and geoscientists, the pathways are somewhat narrower but opportunities exist to move into the ground water industry, environmental consultancies or civil engineering and construction companies.
- **For office and support roles,** the options are much more varied. These roles are, in fact, not sector-specific and it is unlikely that prospective employers will look at working for an oil and gas company any differently to working for other large-scale employers. Office and support roles are vital for any thriving modern economy – whether it be sales, marketing and communications or recruitment, HR and training – companies large and small need these sorts of people in abundance and the experience of working for large oil companies will be invaluable to those who have that experience.
- **And finally, for leadership and management roles,** the same rules apply as for office and support roles. One of the failures of many a local content policy has been the lack of local citizens in senior management roles. It is vital

that these roles are filled by local candidates – at least in part – as this gives those candidates vital experience that they can take into similar roles in other enterprises. Developing local managers and leaders is also likely to drive up the number of new entrepreneurs who, having seen how business works, feel they want to take on that responsibility themselves.

Generally, for those who have worked across the oil and gas sector – particularly those who have been in operational as opposed to office roles – there are a set of common characteristics that are typically developed over time and that any employer from any sector should be glad to have. These include the following:

- Dedication to the job
- The ability to follow logical processes and procedures
- The propensity to be able to adapt to new challenges and to new environments
- The ability to learn new processes quickly and assimilate them into their work
- A commitment to safety and to safe ways of working
- A familiarity with working in multicultural teams

Taken alongside the wealth of technical and vocational skills that an oil and gas industry demands, we can see what potential lies within a workforce that has been truly localised. The challenge – for governments particularly but also for the industry – is to fully understand the nature and scale of the skill base that exists and to then develop coherent plans for moving towards a more diversified economy that draws on these skills and talents whilst at the same time minimising a continued reliance on oil and gas as the drivers of economic stability and employment growth. The current economic climate around oil and gas operations – discussed extensively in Chapter 1 of this book – is testament to the folly of this approach. If a country can capitalise on its skill base and understand the potential for transferability across different sectors of the economy, then the benefits of localisation can be felt far beyond the oil and gas sector.

3.4 HOW LOCALISATION CAN CREATE AN EDUCATION AND TRAINING LEGACY

Allied to the concept of skill transferability is the legacy that localisation leaves in terms of education and training. In our model of localisation, it is not only the workforce and the supply chain that is localised. A critical part of achieving localisation – and a key part of the model – is that the education and training required to support local workforce development and to build supply chain capacity is delivered in country. In the short term, this creates significant residual benefits in terms of bringing greater accessibility to education and training opportunities for local citizens, reducing training costs for operators and generating new jobs and opportunities within the local education and training sector. However, the longer terms benefits – the legacy – are perhaps even more compelling as they provide a country with the means to realise the economic diversification that we have hitherto been discussing.

When we characterise that education and training legacy, there are a number of elements that, if effective localisation is achieved, can provide the springboard for better skill development outcomes across every sector of the economy. These are defined as follow:

- **Institutional capacity**
  The ability of local institutions – colleges, polytechnics and universities – to deliver programmes of learning that are relevant, high quality and commensurate with international standards is a central tenet of the localisation approach. This will only come with institutional capacity building (including a focus on leadership). Once this capacity is fully developed, however, it stays. Through strengthening national institutions for oil and gas programmes, governments and others build a lasting capacity that enables those institutions to thrive whatever the economic demands they are trying to meet.

- **A strong private training sector**
  In any model of localisation, the private training sector is a key part of creating the learning experiences that oil and gas employees need. Training for oil and gas needs to be highly responsive, often technology-led and able to react to the changing demands of the industry. By ensuring that local training providers are supported in meeting these demands, those providers will build their capacity and will be able to diversify into other areas of training provision. As the economic diversification begins to happen, so the private training provision can match and support that diversification.

- **Great teachers, great trainers**
  At the heart of any effective education system are the people who deliver learning. When we think about building capacity within the education and training sector, the first thought is about finding and then developing great educators. It will be an inevitable consequence of localising oil and gas training that you will create a cadre of high quality local teachers and trainers who have the skills and techniques to work with learners across a wide range of subject areas.

- **Facilities that are fit for purpose**
  Alongside teachers and trainers, appropriate equipment and facilities are critical for many oil and gas training experiences, particularly for technical roles. Although there are aspects of oil and gas training that are unique to that industry, much of the equipment used can be useful for the training of related disciplines. More generally, we are increasingly reliant on technology for the delivery of all types of technical and vocational training. Institutions that are able to leverage oil and gas demands as a way of equipping themselves with new technologies will then be well-placed to adapt to the training demands of a diversified economy as and when they need to.

- **A system that is competency based**
  One of the benefits of the oil and gas industry is that it is very demanding in terms of the type of training required and very forward-thinking in relation to pedagogy and models of learning. Competency-based models of learning are now commonplace across the oil and gas industry and are generally

recognised as being the most productive and comprehensive approach to skills and workforce development. A further legacy from localising oil and gas education and training will be the embedding of competency-based models amongst the local institutions, private providers and, in some cases, in relation to the education policy of a particular government. Such a system is easily adaptable to any sector.

- **Coherence between industry and education**
  At the heart of any successful workforce development strategy is the effective connection between industry demand for skills and the supply-side outputs from the education and training system. In our model of localisation, the coherence between what the industry needs and what the education and training system is able to produce is vital. If a country can get this right in the oil and gas industry, the model can then be applied effectively and successfully to other industrial and commercial sectors. In this, oil and gas – and the education and training system that supports it – can become an exemplar for the wider economy.

- **Research capacity**
  Part of the challenge in developing a local education and training system in resource-rich countries is to ensure that the research capacity within knowledge institutions is nurtured and fully supported. Although research is not, in itself, an area that is likely to generate any significant level of direct employment, it is vital that oil- and gas-producing countries invest in their knowledge base and create research hubs that can feed into the thinking around the kind of diversification – both in terms of energy use and wider economic activity – that we have been discussing here. In our model of localisation, investing in higher education institutions will generate significant research capacity that can be exploited as countries seek to diversify.

- **Partnerships and collaborations**
  In every successful model of education and training for oil and gas, partnership and collaboration are key success factors. Within this context, we may think about partnerships between international education and training providers and local institutions. We may also think about partnerships between industry on the one hand and educators on the other. If the oil and gas sector can be a trailblazer in defining and developing these kinds of partnerships, other sectors – and other actors – can learn from this example.

Taken together, these factors clearly indicate how powerful an effectively localised education and training system for oil and gas can be the lever for wider improvements across educating and training as a whole. This powerful legacy adds further weight and impetus to our proposition – that localisation is the only viable and sustainable model of operation for the oil and gas sector in the 21st century.

3.5 CONCLUDING REMARKS

The case for countries – and for companies – to adopt a progressive model of workforce and supply chain localisation has been clearly and passionately made across the pages of this book. If we have done our job, we will have hopefully convinced some of our readers of the value of this model and, furthermore, shone a light on how such a model might be achieved in practice. In this last section of the book, we have tried to explore some of the wider implications of localisation beyond the realm of oil and gas. Not only is this in the interests of ensuring our book does not go out of date as the world transitions to alternative forms of energy. It is more because we have too often seen resource-rich countries squander the wealth and opportunities that hydrocarbon finds promise. And we believe that there are better ways to ensure that a positive legacy remains long after the last well has been capped. However, this legacy will only be secured if those involved in oil and gas production take action today and invest strategically. The implications for both host governments and for operating companies are as follow:

The implications for governments

Securing the kind of legacy we have talked about in this section takes energy, intelligence and a long-term view. The responsibility of governments within this context is to have a vision for how their economic and industrial activity can and should transition away from oil and gas, both vertically – into new forms of energy production – and horizontally into other sectors. This takes coordination, planning and an awareness of the skills, talents and expertise within that country. It also takes proactive policy and legislation to support economic diversification and stimulate inward investment. And it takes a genuine and sustained commitment to investing in education and training at a local level. It is only then that effective localisation of oil and gas operations can be achieved and only then that the legacy from that process will accrue.

The implications for oil and gas companies

Oil and gas companies must recognise that the benefits of localisation are not only financial or related to reputation (although those are critical). For these companies to survive and thrive in a world of fluctuating prices, variable demand and profound shifts in how we think about generating our energy, they too must embrace the concept of diversification and embrace it at a local level in the countries where they operate. By being an active and willing partner in this approach, these companies will position themselves to build new businesses in the emerging markets that, tomorrow, will become the world's economic powerhouses. What's more, with national oil companies taking an ever greater share of the global oil and gas market, international companies need to demonstrate an enthusiasm for developing and embracing new models of collaboration and partnership with host governments and with local communities and citizens. Otherwise, these companies run the risk of becoming the relics of a bygone age of big oil, consigned to history and replaced by something more equitable, progressive and sustainable. The choice is theirs.

Index

Printed in the United States
By Bookmasters